EYES ON ADVENTURE™
EXPLORING

EARTH & BEYOND

Kidsbooks
Incorporated

Copyright © 1998 Kidsbooks Inc.
3535 West Peterson Ave.
Chicago, IL 60659

Visit us at www.kidsbooks.com
Volume discounts available for group purchases.

Table of Contents

Photo Credits: The Sea

Wayne & Karen Brown: pages 10, 17, 22, 28
Michael J. Giudice: pages 9, 29
Francois Gohier: page 17, 26, 27, 29
Tom & Pat Leeson: page 27
NASA: page 8
Norbert Wu: pages 9, 21, 24, 25, 26, 30, 31
Al Giddings Images Inc.: pages 23, 24
Vince Cavataio/Allsport USA: page 14
AP/Wide World Photos: pages 15, 22
AP/National Maritime Museum: page 23
The Bettman Archive: page 12
Gamma Liason: page 13
Kaku Kurita/Gamma Liason: page 31
Pierre Perrin/Gamma Liason: page 29
Gamma Tokyo: page 30
The Granger Collection: pages 12, 13, 22, 26, 31
Doug Perrine: Innerspace Vision: page 10
J.L. Dugast/Liason International: page 8
Lyle Leduc/Liason International: page 12
Carl Schneider/Liaison International: page 15
Hamblin/Liaison International: page 16
Paul Kennedy/Liaison International: page 28
Paul Souders/Liaison International: page 28
Nance Trueworthy/Liaison International: page 16
Brandon Cole/Mo Young Productions: page 16
Mark Conlin/Mo Young Productions: pages 21, 31
Darodents/Pacific Stock: page 8
Reggie David/Pacific Stock: pages 20, 21
Sharon Green/Pacific Stock: page 9
William Bacon/Photo Researchers: page 27
Chesher/Photo Researchers: page 23
David Hardy/Science Photo Library: pages 15, 21
NASA/Science Photo Library: page 14
Hal Beral/Visuals Unlimited: page 11
David B. Fleetham/Visuals Unlimited: page 20
Will Troyer/Visuals Unlimited: page 11
Kevin Deacon/Waterhouse Stock Photography: page 28
Stephen Frink/Waterhouse Stock Photography: pages 18, 19
Marty Snyderman/Waterhouse Stock Photography: pages 10, 25
James Watt/Waterhouse Stock Photography: page 11

Illustrations
Howard S. Friedman: pages 11, 14, 20, 21

Scientific Consultant:
Steven L. Bailey
Curator of Fishes
New England Aquarium

Front Cover: center - NASA; upper left - Gamma-Liason;
upper right - Marc Galindo/Custom Medical Stock
End Pages: front - Wayne & Karen Brown; back - Howard S. Friedman

THE SEA

WATERY PLANET

If you were to rocket into space and look back at Earth, you would see a big blue planet. It looks blue because most of Earth's surface is covered with water. The sea, in fact, has one thousand times more room for living creatures than air and land combined.

GREAT OCEANS

The Pacific, Atlantic, Indian, and Arctic oceans cover over two-thirds of our planet. That's more than 300 million square miles! The Pacific is by far the deepest and the largest ocean. It covers more than one-third of the globe.

PASS THE SALT

What's the saltiest ocean? The Atlantic. Rocks make the water salty. Waves erode the rocks, which contain salt that dissolves in the water. Some sea animals can't live in very salty waters. Others, like clams and oysters, use the calcium found in sea salt to build their shells.

Salt literally drips from rocks near the Red Sea—one of the saltiest seas on the planet.

WATER RE-CYCLE

How do Earth's oceans remain so full of water? The answer is in the water cycle. When water evaporates from the sea it becomes rain clouds. When the rain falls onto land, it drains into rivers. Then the rivers take water back to the sea.

DOWN UNDER

Home to many fascinating creatures, and burial ground for countless shipwrecks, the sea is a gold mine to adventurous types. Divers investigate animal behavior, recover ancient wrecks, and discover new marine species.

A diver observes a giant seajelly in action. ▼

SEA STUDY

Scientists who study the oceans are called *marine biologists*. They try to figure out how and why the oceans change, and why certain sea creatures and plants live in one place and not in another.

FIRE AND ICE

Close to the equator, where the climate is hotter than anywhere else on Earth, sea water is warm. Farthest from the equator lies the Arctic Ocean, home of glaciers, icebergs, and animals specially adapted to wintry weather. Here the sea is icy cold.

NOW AND THEN

About 6,000 years ago the ancient Egyptians invented sails. Over the centuries, sailing became a vital means of transportation and industry. Today, sailing is a pastime and a sport. One of the world's most famous ocean sailing races is the America's Cup.

9

WHO LIVES HERE?

Try to imagine all the living things in the world—more than 10 million species of animals, plants, fungi, bacteria, and other types of creatures! Does it come as a surprise that only 20 percent live on land? The remaining 80 percent are found in the sea.

◀ The octopus was once thought to be a monster.

▲ The hermit crab scuttles across the sea floor.

SINK OR SWIM

When you think of water, you probably think of swimming. But not all sea creatures spend their life swishing their fins like fishes or whales do. Many plants and animals live on the sea floor. Tiny plants and animals known as plankton simply float on the ocean currents.

◀ The wide-mouthed manta ray easily gobbles up some plankton.

UP FOR AIR

Sea mammals, such as whales and dolphins, cannot spend all of their time under the water's surface like fish. They must come up to breathe, as people must do when in the water. However, sea mammals can dive for long periods of time. The sperm whale can stay underwater for more than an hour, holding its breath while hunting giant squid.

HOMEBODIES

Many sea creatures stay in one area of an ocean their whole life. Certain animals, such as the giant whale shark at right, roam the waters for food.

Hatched in rivers, salmon live in the ocean during their adult life. But when it comes time to spawn, or produce young, salmon leave the ocean and swim back to the river where they were born.

STAYING ALIVE

As a lower link on the food chain, small fish have developed a great defense—swimming in schools. Because the fish swim together, darting left and right, predators have a hard time picking out a single fish to catch.

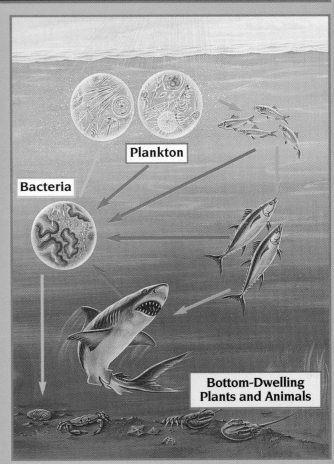

Plankton

Bacteria

Bottom-Dwelling Plants and Animals

THE FOOD CHAIN

Like any chain, a food chain is made of links—living creatures eating other living things. It all starts with bacteria, which is partly dependent on the decomposition of dead animals. Bacteria provides nutrients to plankton and other sea life. Then plankton are eaten by small animals who are in turn eaten by larger animals.

11

ALL ABOARD

Boats have been transporting people and goods for thousands of years. By the 20th century, advances in technology made ocean travel faster, more reliable, and more comfortable.

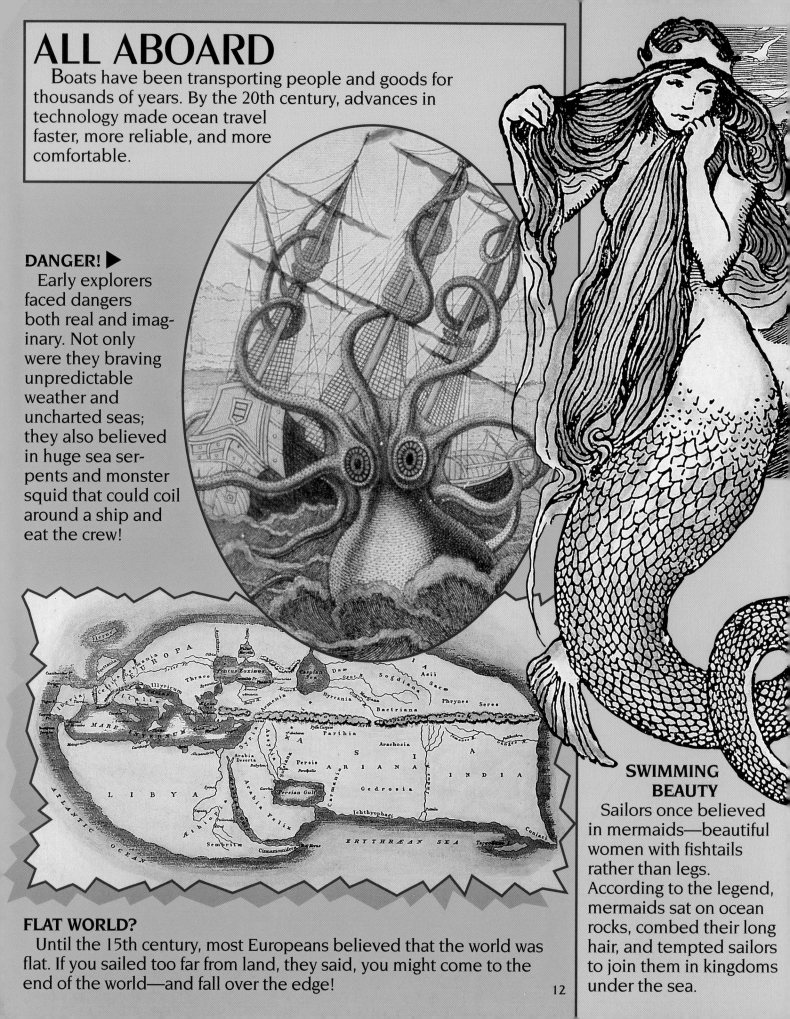

DANGER! ▶

Early explorers faced dangers both real and imaginary. Not only were they braving unpredictable weather and uncharted seas; they also believed in huge sea serpents and monster squid that could coil around a ship and eat the crew!

FLAT WORLD?

Until the 15th century, most Europeans believed that the world was flat. If you sailed too far from land, they said, you might come to the end of the world—and fall over the edge!

SWIMMING BEAUTY

Sailors once believed in mermaids—beautiful women with fishtails rather than legs. According to the legend, mermaids sat on ocean rocks, combed their long hair, and tempted sailors to join them in kingdoms under the sea.

VIKING MIGHT

Some of the earliest explorers were Vikings. Eric the Red sailed all the way from Norway to Greenland around A.D. 980. His son Leif Ericson sailed even further—from Iceland to the east coast of Canada! This was around A.D. 1000—nearly 500 years before Columbus.

AROUND THE WORLD

Ferdinand Magellan is credited with circling the globe and proving the world was round, but it was his navigator, Juan Sebastian del Cano, who captained the ship that finished the voyage in 1522. Magellan was killed in a battle the year before.

DEEP-SEA VESSELS

For centuries, people traveled *on* the water. Not until the 18th century, with the invention of the submarine, was *underwater* travel made possible. From the submarine came submersibles—small submarines used for marine research, archaeological expeditions, and pleasure rides.

▼ Today, ocean travel can be a vacation. Cruise ships offer luxury and spectacular views.

13

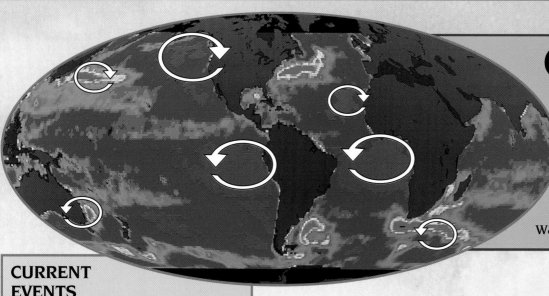

ON THE MOVE

The ocean stays in motion. Even when the water seems calm—waves, tides, and currents keep the waters flowing.

CURRENT EVENTS

Not until the 20th century did people begin to understand the ocean's currents. In the Northern Hemisphere, currents sweep clockwise from the equator to the Arctic. In the Southern Hemisphere, they travel counterclockwise.

TIME AND TIDE

Tides are caused by the sun and moon's gravitational pull. As Earth turns on its axis, one side of the planet comes closer to the moon. Pulled by the moon's gravity, the ocean may rise toward the moon and become deeper, which is called high tide. On the opposite side of Earth, the ocean may be more shallow—called low tide.

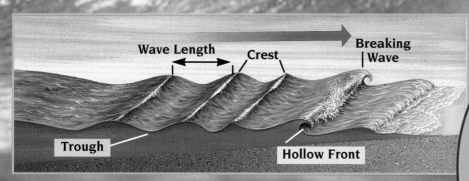

Wave Length Crest Breaking Wave

Trough Hollow Front

MAKING WAVES

How do waves form? Wind blowing along the sea's surface drags the top of the water and creates waves. As a wave moves closer to shore, the water below the wave gets shallow. This makes the wave taller until it moves into even shallower water, topples over, and breaks.

BIG BOARDS

Surfing has been an exciting sport since A.D. 400, when it was invented by the Polynesians in the Pacific. The first Westerner to witness surfing was Captain James Cook at Kealakekua Bay, Hawaii, in 1778. The surfboards, shaped from trees, were 20 feet long and weighed as much as 200 pounds!

◀ KILLER WAVES

Tidal waves, or *tsunamis*, don't actually have any connection to tides. They are caused by seismic activity on the sea floor. Tsunamis can travel hundreds of miles, growing sometimes hundreds of feet high, and reaching speeds of 500 mph.

WATER SPORTS ▼

A combination surfboard and sail, "sailboards" are speedy and acrobatic. Some windsurfers have been clocked at 50 mph, moving faster than any other sailing craft!

▲ KON-TIKI EXPEDITION

In 1947, explorer Thor Heyerdahl built a simple wooden ship similar to the one used by early peoples. He wanted to test whether Indians from South America could have sailed the seas to settle Polynesia. It took him 101 days to sail from Peru to Polynesia, which proved the early expedition possible.

MEETING THE LAND

The coast can be a loud and dramatic place, where rocky cliffs are hit hard by breaking waves. Coastal creatures have to cling to rocks to keep from washing out to sea. Barnacles, snails, and bivalves—such as mussels and some types of scallops—attach themselves to coastal rocks with a sticky secretion produced in their body.

PUDDLE DWELLERS

When the tide goes out, puddles of water are left on shore. Seastars, seaweed, periwinkle snails, crabs, and sea urchins make their homes in these tide pools.

SHELL TREASURES

Have you ever found a queen conch shell on the beach? How about a tiger cowry or a rosy harp? These beached shells were once the homes of soft-bodied animals known as shellfish, or mollusks. Empty shells are popular collectible items for beachcombers.

WEED FOREST

Kelp, or seaweed, grows in "forests," taking root on hard sea floors near shore. The fastest-growing plant, kelp can gain as much as three feet a day, and reach lengths of 213 feet! Kelp floats upward instead of sinking because its stems are surrounded by gas-filled bulbs.

OTTER HANG-OUT

Kelp forests serve many purposes for the sea otter. During the day, otters eat the abalone, crabs, and urchins that feed there. At night they wrap themselves in the kelp fronds, and the waves rock them to sleep.

CITY OF CORAL

One of the most colorful and populated areas of the sea is in and around a coral reef. The bright colors of the coral are caused by algae that live inside. Outside the coral, schools of tropical fishes dazzle the eye with an equally fascinating display of color.

NO TOUCHING!

One of the best ways to appreciate the living treasures of a coral reef is to see it up close. Snorkelers and scuba divers flock to reefs. But they are careful never to touch the coral because that can damage or kill it.

IT'S ALIVE!

Coral may look and feel stony, but it is not rock. It's the skeleton of a living animal called a polyp. The polyps grow a skeleton on the outside to protect and support their soft bodies. Because the reef-building corals cannot live in water colder than 64°F, they are found only in warmer waters.

FEEDING TIME

All animals have to eat, including coral polyps. How do they do it when they are attached to the ocean floor? Coral polyps actually have tiny arms that catch plankton and pass it into their mouth.

THREE REFS

Reefs grow in different ways. A *fringing* reef is attached to the shore. An *atoll*, like the one shown at left, is a ring of coral formed around a sunken volcano. A *barrier* reef has a channel of water between it and the shore. Australia's Great Barrier Reef is a whopping 1,250 miles long. That makes it the biggest structure ever built by animals.

COAT OF ARMOR

You can often count on finding clownfish with the grasslike sea anemone. Unlike most sea life, the clownfish is safe from the anemone's stinging cells because of a thick, slimy mucous on its body.

Brain coral

KNOW YOUR CORAL

Stony corals, or hard corals, such as brain corals, form reefs. Gorgonians, or soft corals, such as sea fans, grow on the sea floor and on reefs, and look a lot like ferns or bushes.

CORAL POPULATION

Every coral reef has a population consisting of thousands of different animals that live and thrive there—including shellfish, moray eels, sea horses, and sharks.

SEASCAPES
Under the ocean there is a landscape almost as varied as the one above water—with gigantic canyons, plains, mountains, and caves.

AGE-OLD SEA
When plants and animals die in the ocean, their remains drift down to a sea bottom. Drilling deep with special tools, scientists take samples of this sediment. The samples provide a rich history of millions of years of ocean life.

At the center of the Atlantic Ocean is a mountainous ridge surrounded by plains and valleys.

North America

Europe

South America

Africa

UNDERWATER MOUNTAIN
Most islands are really the peaks of underwater volcanic mountains. Movements in the earth's crust can produce heat and pressure inside an underwater volcano. The pressure eventually causes the volcano to "blow its top." Lava, dust, and rocks flow out, covering the volcano layer by layer, until it breaks the ocean surface and makes an island.

Sea Level

Crust

Mantle

ECHO MEASURE
To measure the ocean's depth, scientists transmit sonar (sound) pulses toward the sea floor, then listen for the echo. The longer it takes to hear the echo, the deeper the water.

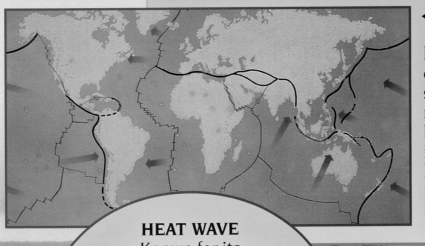

◀ TRENCH VALLEY

Beneath the seawater and land are pieces of the earth's crust known as plates. In some places on the seafloor, mountains and valleys have been formed by movements in the plates. The largest valley is the Mariana Trench, near the Philippine Islands. It is almost 7 miles deep.

HEAT WAVE

Known for its earthquakes and volcanic eruptions, the "Ring of Fire" is a 24,000-mile circle of volcanoes in the Pacific Ocean. (Above, the "ring" and other volcanoes are represented by small orange dots.)

← **Volcano**

← **Shallow Magma Chamber**

← **Rising Magma**

Deep Source of Magma

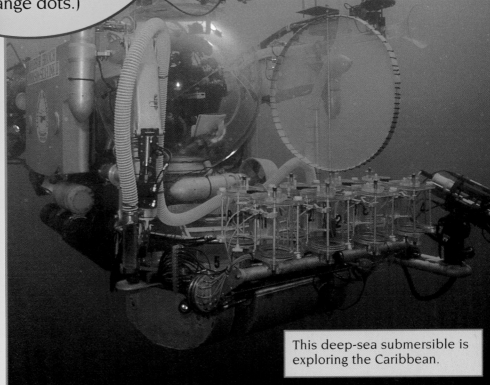

This deep-sea submersible is exploring the Caribbean.

FAMOUS PROJECT

When submersibles were used to explore the Ring of Fire in 1974, scientists discovered huge rock chimneys venting clouds of scalding hot water!

21

TO THE BOTTOM

In previous centuries, one danger to seagoing merchants were pirates, who raided ships the world over. A great deal of loot was lost to the sea bottom. No one could dive into high-pressured, deep waters to retrieve the treasures. Today, the right equipment can get an explorer almost anywhere.

About 170 feet below the surface of the sea, scientists explored a 3,400-year-old shipwreck. Among the finds: pottery, bronze weapons, and gold. ▶

HARD HATS

In 1819 Augustus Siebe invented a copper diving helmet (weighing 20 pounds) that allowed divers to reach depths of 200 feet. A long hose, which stretched from the helmet to a pump on the surface, brought air to the diver.

AQUA LUNG

Scuba divers wear tanks containing air, which is fed into the diver's mouthpiece. This breathing device, called the *Aqua Lung*, was developed by Jacques Cousteau, the famous French oceanographer, and Emile Gagnan in 1943. It allows divers to explore the sea as deep as 500 feet.

▲ BRAVE MR. BEEBE

American explorer Charles Beebe was the first to descend to depths no diver could reach. He designed a spherical steel vessel called a bathyscape that could be lowered from a ship by a long cable. In 1934 he reached a depth of 3,028 feet!

DEEP BREATHS

In the deep sea, explorers must wear jointed metal suits that are heavy enough to withstand great water pressure. This kind of atmospheric suit makes it possible for a diver to walk on the sea bottom 2,000 feet below the surface. At right, the submersible *Star* II and the atmospheric diving suit known as JIM take divers to the bottom.

◀SEA LAB

Because underwater living would provide a unique view of the sea, a number of scientists have become *aquanauts*, forsaking land for a few weeks to be with the fishes. With the help of atmospheric suits, vehicles, and robots, the sea may become known in the future as the marine biologist's laboratory.

RECORD DEPTHS

Swiss physicist Auguste Piccard modified the bathyscape so that it could go even deeper. His son Jacques set the record in 1960, when he explored the Mariana Trench, 35,800 feet, almost 7 miles under the sea.

ROBOTIC EXPLORERS

Today, robot submersibles, which can descend 12,000 feet, are used to salvage treasure and explore the sea. In 1985 the *Argo* located the shipwrecked remains of the great ocean liner the *Titanic* (above), which sank in 1912. Its brother robot, *Jason*, was used to explore the wreck.

23

MYSTERIES OF THE DEEP

In 1872, the *Challenger* expedition of British oceanographers proved beyond a doubt that there was life deep down in the ocean. Using scoops and dredges attached to ropes, they gathered samples of 4,417 new marine organisms! But scientists have only recently begun to fathom the mysteries of the deep.

◄ The gulper fish has a huge mouth for swallowing large prey.

▼ This hatchetfish is being pursued by the viperfish (lower right), another deep-sea creature with light organs.

CLIFF HANGER

At a certain distance from each continent, the ocean floor drops sharply to a depth of 20,000 feet. In the very deep sea there is no sunlight, no plants, and the water is icy cold. Below a depth of 7,000 feet in any ocean, the temperature never rises above 39°F!

VOLCANIC CREATURES

Deep down on the ocean floor are vents that spew out scalding hot water. Warmed by liquid rock inside the earth, these springs are rich in minerals. Giant clams, tube worms twelve feet long, and blind crabs and shrimp the size of small dogs, all live near hot-water vents. They eat a special bacteria that manufactures its own food from the vent's gases and heat.

◄ NIGHT VISION

Fish in the deep sea are specially adapted to the darkness in which they live. The anglerfish, which glows in the dark, has its own "rod and bait." On the rod are lights that attract prey. A second set of teeth in the back of its throat prevent prey from escaping.

Giant tube worms

DO NOT DISTURB ▶

In 1938, fishermen in the Indian Ocean netted a coelacanth, a fish believed to have been extinct for 100 million years! Scientists speculated that it had been living undisturbed in the deep sea.

UNDER PRESSURE

Animals living in the deep ocean have adapted to the tremendous pressure of the water. Most are so perfectly adapted to this environment that they cannot survive for long when brought up to the surface—the change in pressure is just too much.

BON APPETIT ▼

Did you ever wonder who keeps the ocean floor clean? Sea cucumbers help by eating the muddy surface to digest what little food it contains.

POLAR WATERS

Have you ever been on top of the world? How about the bottom? They are pretty cold places. The North and South poles, or the Arctic and Antarctica, are known for their icebergs, glaciers, extreme weather, and polar wildlife. They also have a reputation for being dangerous to explorers!

BIG FREEZE

At any one time, there are about 200,000 icebergs floating in the Antarctic Ocean. They look big up top, but 90 percent of their volume is actually found beneath the ocean's surface. Scientists think all this water could be put to good use—to irrigate drought-stricken land, for example. But there's one problem. How do you move an iceberg?

NORTHWEST PASSAGE ▲

For hundreds of years, explorers tried to find a route along the northern coast of Canada which would link the Atlantic to the Pacific Ocean. Searching for the Northwest Passage was dangerous. Ships became frozen into the ice, or were wrecked on icebergs. A route was finally discovered in the 1850s.

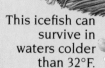

This icefish can survive in waters colder than 32°F.

KEEPING WARM

What would you do to keep warm if you lived in a polar sea? Icefish, and some other polar animals, have special chemicals in their blood that prevent their body's fluids from freezing. Mammals such as whales, seals, and walruses have layers of fat, called blubber, to insulate them from harsh temperatures.

The beluga whale lives in shallow Arctic coastal waters. ▶

A KNACK FOR KAYAK

The Inuits, or Eskimos, developed the kayak—a variation on the canoe, constructed from driftwood and stretched sealskins—to hunt walrus and other Arctic animals. Today's kayaks are fiberglass, and they are paddled all over the world, not just in the Arctic!

POLAR POPULATION

Penguins live in the Antarctic. Polar bears and walruses live in the Arctic. Seals, orcas, and most whales live at both poles. In the winter, seals and walruses hunt under the ice. Seals make holes in the ice where they come up for air.

27

BOUNTY OF THE SEA

Treasures in the sea are not limited to those lost in shipwrecks. The ocean is a source for food, energy, and valuable minerals.

◀ Salt has been mined from the sea since ancient times.

HOMEMADE JEWELS

What happens when a grain of sand finds its way into a mollusk, such as an oyster or a conch? In one mollusk out of ten thousand, the sand becomes coated with layers of *nacre*, the stuff that forms the shell's shiny insides.

Several years later a pearl is formed! To improve the odds of getting more perfect pearls, people culture the gems themselves, implanting a "seed" in an oyster around which a pearl may grow.

OIL BELOW ▶

Offshore oil wells supply about 17 percent of the world's petroleum. Most rigs are in shallow water, but deep-sea drilling techniques are being developed that could double or triple world production.

HYDRO HEAT

Looking for new energy sources, scientists are developing a new process using seawater. The heat absorbed from the sun could be stored in the water and converted into electricity.

▲ WHAT A CATCH!

A modern fisher's catch consists of fishes that swim in schools, such as tuna, salmon, anchovies, and sardines; fishes that keep to the sea floor, such as cod, haddock, and flounder; and shellfish that are harvested from shallow waters, such as oysters, clams, scallops, and lobsters.

WHALE WATCHING

In the 19th century, more than 70,000 Americans worked in the whaling industry. Whales were hunted to near extinction for their oil and other parts. Whale watching, a new popular million-dollar industry, encourages people to think ecologically.

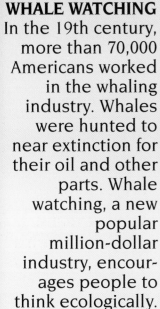

Fishers break holes in the ice to hook passing fish.

GONE FISHING

In prehistoric times, fishers used sharpened bones for hooks, and vines for fishing line. Modern fishing fleets are like floating factories. Huge trawlers can haul in tons of fishes with each giant net. The catch is then cleaned and quick-frozen at sea.

In Mexico, these fishermen cast a net for their catch.

ENDANGERED OCEAN

All the waters of the world are connected. If one sea, or even one link in the food chain is damaged, people will also be affected. For that reason, pollution could be the real-life sea monster of the modern age.

Sewage dumped along coastlines can produce deadly bacteria in our drinking water and seafood.

DISASTER!

Oil and water don't mix, so an oil spill can spread and cover hundreds of square miles. When the *Exxon Valdez* tanker ran aground in Alaska in 1989, dumping 10 million gallons of oil, it killed hundreds of thousands of marine animals.

Workers attempt to clean up ▶ the oil spilled by a tanker.

POISONED WATER

Rain washes pesticides and fertilizers from farmers' fields and homeowners' lawns into the sea, causing algae to grow out of control and destroy other living things, such as turtles and shellfish. Poisonous emissions from factory chimneys travel through the air and fall upon the land or sea as "acid rain."

CLEAN IT UP!

What's to be done about pollution in the ocean? Enforcing laws will help curb carelessness. The Clean Water Act of 1977 mandates controls and clean-ups for industrial and municipal pollution. Federal Safety Regulations, imposed by the U.S. Congress after the *Exxon Valdez* oil spill, assign the cost of clean-up to oil companies.

FISH FARM
Due to advances in technology, the ocean is in danger of being over-fished. The solution may be *aquaculture*—fish farming. After they hatch from eggs, the fish are fattened up in pens until "harvest time."

The unintentional killing of other sea creatures during shrimping adds to the problem of overfishing the ocean.

A modern fish farm

SAVE THE DOLPHINS!
The fishing nets used to trap tuna often catch dolphins and other marine life, such as the shark shown above. Concern over the many dolphins killed caused major tuna canneries to stop buying tuna caught in nets. Ecologically-minded fishers use hooks instead.

MAN OF THE SEA
By exploring, scientists are learning more about the sea and the marine life it supports. One of the greatest ocean explorers was Jacques Cousteau (1910-1997). He made award-winning documentaries and wrote many books about his discoveries so that others could experience the beauty of this vast underwater realm.

NATURAL DISASTERS

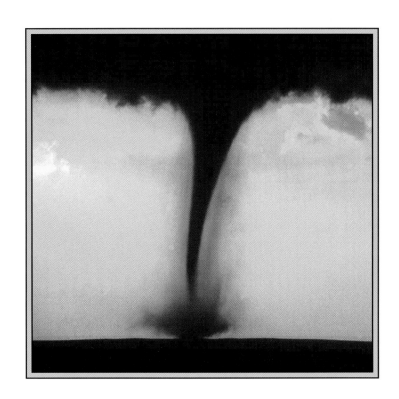

IN NATURE'S PATH

Earthquakes, tornadoes, floods—these are some of the severe acts of nature that affect our planet. What exactly makes them disasters? They are destructive, causing untold hardship and claiming thousands of lives every year.

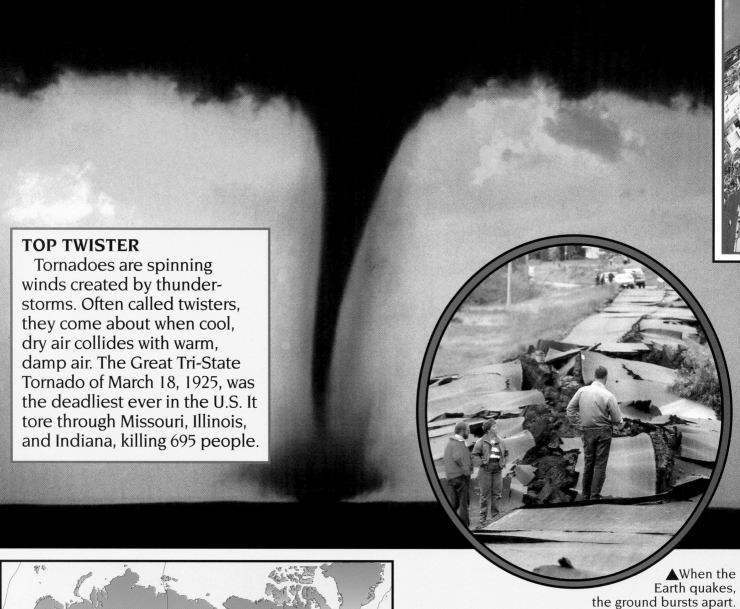

TOP TWISTER

Tornadoes are spinning winds created by thunderstorms. Often called twisters, they come about when cool, dry air collides with warm, damp air. The Great Tri-State Tornado of March 18, 1925, was the deadliest ever in the U.S. It tore through Missouri, Illinois, and Indiana, killing 695 people.

▲When the Earth quakes, the ground bursts apart.

Approximate location of Earth's plates and the Ring of Fire.

◀ SHAKY GROUND

Both earthquakes and volcanoes are part of the Earth's behavior. Our planet's surface is a rocky crust made up of a dozen or so plates. The plates move, sometimes causing such pressure that the land quakes, especially in an area of the Pacific known as the Ring of Fire. Other times, liquid rock can come up through spaces between the plates, forming a volcano.

SNOWBOUND ▶

A big snow, or blizzard, can cause a lot of damage and can shut down a city. However, snow can be fun, and some people make the most of it.

◀ $$$

Hurricanes are formed from tropical storms. Unlike tornadoes, the winds of a hurricane can destroy a broad area and can last for days. Hurricane Andrew was one of the costliest in U.S. history, causing over $46 billion in damage. The storm left over 258,000 people homeless and tore through much of the Everglades National Park.

▼ BURN UP

Dry weather and soaring temperatures often make conditions in the western United States perfect for forest fires.

TO THE RESCUE ▶

Is there anything reassuring in the face of natural disasters? There is often a warning. Forecasters have modern technology at their fingertips—computers, satellites, radar, and more. Also, when a disaster does strike, people come together and help each other out, and professionals come to the rescue—firefighters, medical workers, and special disaster teams.

◀ RAGING RIVER

When it rains, and it pours, it sometimes floods. Imagine two months of rain, and the mighty Mississippi River overflowing its banks. In 1993, nine states in the Midwest experienced the costliest flood in U.S. history —$20 billion in damage was done and about 70,000 people were left homeless.

35

EXPLODING GIANTS

Volcanoes are sometimes called sleeping giants. They can rest quietly for decades or even centuries. But when they wake up, watch out!

An eruption of ▶ Kilauea in Hawaii.

OVER THE TOP

The deadliest volcano ever to erupt blew its stack in Sumbawa, Indonesia, in April 1815. Lava and gases killed 92,000 people.

▲ A volcanic ball of fire.

BLAST FROM THE PAST!

One of the biggest volcanic explosions ever recorded took place on an island in Indonesia. On August 27, 1883, Krakatau volcano erupted, killing more than 36,000 people and destroying about 160 villages. It is estimated that the explosion was over 25 times more powerful than the largest hydrogen bomb ever tested!

HOT STUFF ▶

Hot liquid rock erupts from inside the Earth. There, in a layer known as the mantle, temperatures reach a roaring 2,200°F! The thick flowing rock substance is known as magma. Once magma comes out of the ground, it's called lava.

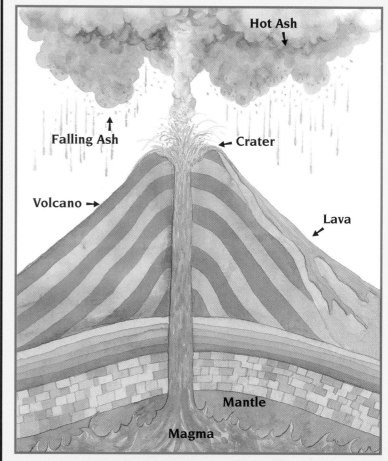

Hot Ash

Falling Ash

Crater

Volcano →

Lava

Mantle

Magma

MOUNTAIN TOPPER ▼
In this infrared aerial photograph taken from 15,000 feet by NASA, you can see the 1980 eruption of Mt. St. Helens. The volcano lost 1,300 feet off its top and caused 57 deaths—more than any other eruption in U.S. history.

CHECKING IT OUT ▼
Volcanologists are people who study volcanoes. They look for signs indicating the next eruption. They also venture out to study volcanoes in action and lava on the move.

SAVED BY THE CELL
About 30,000 people died in the eruption of Mt. Pelée on the island of Martinique in the West Indies in 1902. But many people survived. One was spared because the thick walls of his jail cell protected him from the blast.

▼ LAVA WILL TRAVEL
Lava stopped short of these homes near Mt. Etna in Sicily, Italy, and spared them from complete ruin. But lava can really move, swallowing whole cities and destroying all in its path.

▲ These Hollywood cameramen happened to be nearby when a volcano in Mexico erupted in 1943.

GOOD DEEDS
Although volcanoes do damage, they also cause soil to become more fertile and create entire islands, such as Hawaii and Iceland. Also, there are useful volcanic materials, such as pumice and basalt.

37

LOST CITIES

Almost 2,000 years ago in a Roman city known as Pompeii, a great disaster was about to happen. On the outskirts of town was Mt. Vesuvius. Many people lived near this lovely volcanic mountain. Some farmed its land. Little did they know that the volcano was ready to blow its top!

▲ BLOW UP

On August 24, A.D. 79, Mt. Vesuvius erupted. Within hours, Pompeii was buried under 6 to 20 feet of ash and spongy rock, called pumice. Forgotten, Herculaneum was destroyed by a super hot river of lava. Both cities were not to be discovered for many centuries.

▼ STOPPED DEAD IN THEIR TRACKS

People from Pompeii were buried as they tried to escape. Over the years as they deteriorated, the bodies left a space in the hardened volcanic rock. Archaeologists (scientists who study past cultures) discovered these spaces and poured liquid plaster down into the rock. They later dug out the hardened plaster, creating a permanent record of the fallen victims.

◄ VIEW OF VESUVIUS

Vesuvius has been painted by many artists. This painting depicts people fleeing Pompeii. Citizens who actually survived the rain of volcanic ash returned to their city to scavenge for riches. In later centuries people continued to pillage, and many archaeological treasures ended up in museums.

▲ A house in Pompeii.

◄ In Pompeii, the ovens from a baker's shop survived the volcanic eruption.

LOST AND FOUND

Houses and other buildings have been uncovered in Pompeii and Herculaneum, leaving a perfect time capsule of daily life almost 2,000 years ago. However, much remains to be excavated in Herculaneum. Work there has been more difficult than at Pompeii, because Herculaneum was buried in lava, which is much heavier than ash and pumice.

▲ The house of Poseidon in Herculaneum contains the best-known mosaic discovered in the ruins.

LOOKS CAN FOOL

Look out! Mt. Vesuvius can erupt again. It is, after all, an active volcano. An active volcano is one that is erupting or has erupted since the time that written records have been kept. The last really spectacular eruption of Mt. Vesuvius occurred in 1944, and in 1906 the volcano blew off the ring of its crater.

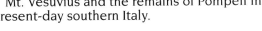

◄ Mt. Vesuvius and the remains of Pompeii in present-day southern Italy.

QUAKES

Like much of nature's fury, earthquakes are short-lived. But their destruction echoes for years. Earthquakes have been taking place since the beginning of time. And although scientists are better able to predict them today, earthquakes still cause enormous damage and great loss of life.

The roof was taken off this house during the 1994 earthquake in Los Angeles.

PURE DESTRUCTION

The Los Angeles earthquake of January 17, 1994, was the most destructive in U.S. history, killing 61 people and injuring more than 8,000. The loss in dollars was close to $20 billion.

DEFINITELY DEADLY
Fortunately, some earthquakes hit where few people live and work. Others, however, offer no such mercy. One of these deadly quakes occurred in Tangshan, China, in 1976. Registering 8.2 on the Richter scale, the quake killed 242,000 people!

▲ Workers rescued this man from a collapsed building after the 1994 quake in California.

DOUBLE TEAMING

The strongest earthquake in U.S. history struck William Sound, Alaska, on March 27, 1964. It measured 8.4 on the Richter scale. Following the quake came a giant wave from the sea. Called a tsunami (sue-NAH-me), the wave traveled at 450 mph and destroyed the town of Kodiak.

This geologist is examining a seismograph after an earthquake measuring 5.0 hit Los Angeles. The machine records movements in the Earth.

◀ A quake does more than shake the Earth. It breaks open water lines and brings down power lines, causing fires and floods.

HELPING HANDS

After the disastrous earthquake in Armenia on December 7, 1988, was reported on television, rescue workers from all over the world came to help. The quake, measuring 6.9 on the Richter scale, killed over 25,000 people, injured 15,000, and left 400,000 homeless.

TAKING MEASURE

Charles F. Richter invented what is known as the Richter scale. It has been used since 1935 to measure the strength of an earthquake. Very few earthquakes register above an 8.0. If one does, it means there's a whole lot of shaking going on!

JAPAN'S HORROR

Located on the Ring of Fire, where four out of five earthquakes occur, Japan gets hit hard by quakes. On January 17, 1995, a quake struck Kobe with a magnitude of 7.2 on the Richter scale, causing about $100 billion in damage.

◀ On September 1, 1923, an earthquake struck the Kanto Plain in Japan, leaving over 550,000 dwellings destroyed.

41

SHAKY SAN FRANCISCO

San Francisco, California, is a fun place to visit. There are cable cars and neat hills, a blue bay, and, beyond it, the Pacific Ocean. But, unless you like living on the edge, you may not want to live there. Sometimes tremors occur and there's little destruction. But when a big quake hits, the effects can be disastrous.

▶ Between quakes, the city enjoys a quiet, peaceful time.

▼ Search teams use dogs to find people trapped in the rubble of collapsed buildings.

The San Andreas fault lines.

Fires following the 1906 quake roared out of control, as the city's main water supply was cut off.

A MAJOR FAULT

San Francisco sits on the San Andreas fault—a 600-mile stretch where a crack exists in the Earth's crust. When the huge plates that make up the crust push into each other, and the colliding rocks can no longer bend, the Earth begins to tremble under the pressure and an earthquake occurs.

CITY IN CHAOS

In the San Francisco earthquake of April 18, 1906, more than 500 people died and over four square miles of buildings were destroyed.

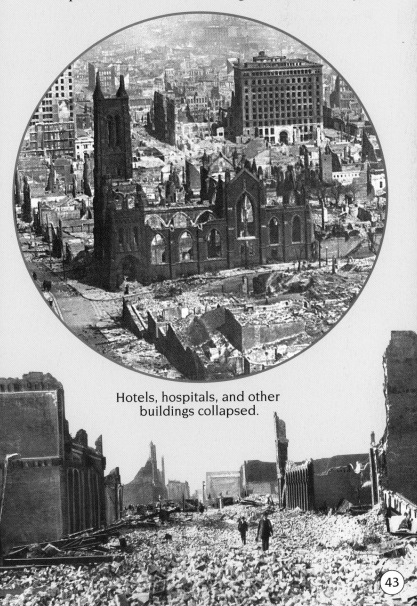

Hotels, hospitals, and other buildings collapsed.

TWISTERS

They spin and race along the ground at speeds of up to 250 miles per hour, picking up anything in their path, including cars and trains. Tornadoes have even been known to pluck chickens, leaving the poor creatures featherless but unharmed.

WHAT IS IT?

It's a twister, a cyclone, a whirl-wind! All these words are used to name a tornado, the storm with the fastest and strongest winds on Earth. But what *is* a tornado? It's a violently rotating column of air produced by a thunderstorm. Funnel shaped, the vortex usually moves over the land in a narrow path, staying in contact with both the thundercloud and the ground.

When a twister contacts water instead of the ground, it forms a waterspout.

ONSTAGE

Between April 3 and 4, in 1974, over 140 tornadoes touched down in 13 states and Canada. It was a lucky day, however, for some drama students at the local high school in Xenia, Ohio. They escaped to a hallway just in time—before two buses landed onstage.

◀ Destruction to Xenia

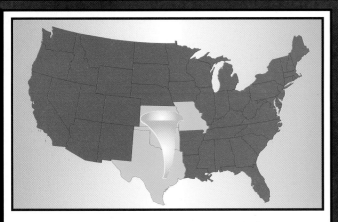

Most tornadoes in the U.S. occur in the Midwest. Kansas, Missouri, Oklahoma, and Texas make up a region known as Tornado Alley. Oklahoma is struck by tornadoes more than any other place on Earth.

LISTEN UP

Tornado season occurs in both spring and fall. If heavy thunderstorms are in your area, turn on your TV or radio. A tornado *watch* means that the thunderstorms could contain tornadoes. A tornado *warning* means that a tornado has actually been detected on radar or seen by people.

WHAT TO DO WHEN A TWISTER HITS

How do you stay safe when a tornado is hurling a barn from its foundation or stripping the roof off a house? Some people have fiberglass shelters buried in their backyards. Others take refuge in the basement.

OFF THE CHARTS

Tornadoes are rated from F0 to F5 based on the damage they do. A rating of F0 means that the damage was light. Maybe some windows were shattered. A rating of F5 means that the damage was "incredible," with houses and cars carried away. The scale, called the Fujita scale, was devised by T. Theodore Fujita, a physics professor.

An F2 can rip the roof off a house.

HURRICANES

They kill more people than any other storm. They uproot trees, flatten buildings, and cause floods. With winds of up to 220 mph, they devastate entire cities. Andrew, Camille, Agnes, and Gilbert—these are the names given to some of the most devastating hurricanes in history.

High tides caused by Hurricane Marilyn ▶ carried boats onto the streets in the Virgin Islands.

Much of the damage from hurricanes is caused by huge ocean waves blown to shore.

1954

It was not a good year for people living along the Eastern seaboard. In August, Hurricane Carol pounded the area. In October, Hurricane Hazel tore through the country, hitting Canada as well.

WHAT'S IN A NAME?

Hurricanes used to have only female names. But since 1979, hurricanes have been named for men and women. The first man's name to be used was Bob.

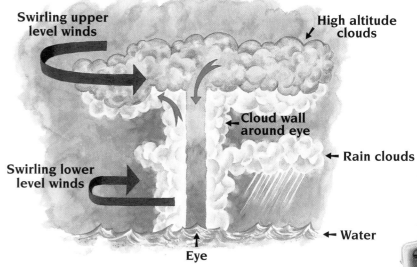

Swirling upper level winds

High altitude clouds

Cloud wall around eye

Rain clouds

Swirling lower level winds

Water

Eye

▲ BIG WIND

Hurricanes form over the ocean when warm air combines with cool air to create wind. If water is present and the wind is strong, a tropical storm forms. About half these storms become hurricanes.

INSIDE THE EYE ▶

A hurricane has a calm center, called an eye, which you can see in this satellite image of Hurricane Gilbert. Satellites help forecasters give early warning to residents and, therefore, help save lives. But satellites weren't around in 1900, when a hurricane hit Galveston, Texas, and killed 6,000. It was the worst hurricane disaster in the U.S.

FAST AND FURIOUS

In 1969, winds of Hurricane Camille reached 200 mph—the fastest sustained winds in a U.S. hurricane—as they pounded Mississippi and Louisiana.

▼ ALSO KNOWN AS

Hurricanes are called typhoons in the China seas and cyclones in the Indian Ocean. Regardless, they're usually deadly. Typhoon Ike devastated the Southern Philippines on September 2, 1984. More than 1,300 people died in flash floods triggered by the storm.

A ▶ U.S. soldier helps a young girl and her family get emergency supplies after Hurricane Andrew destroyed their home in Florida in 1992.

▲ Wreckage from Hurricane Hugo.

IT'S SNOWING!

Snowflakes cover trees, lawns, buildings, cars, and roads with a clean-looking velvety blanket. What could be more enchanting? But snow can also be dangerous. When a light dusting turns into a storm, which turns into a blizzard, few people are thinking about beauty.

BLANKET OF WHITE

The most disastrous winter storm in U.S. history took place in 1888. For about four days, a blizzard dropped five feet of snow all along the East Coast. Over 400 people died, and damage was estimated at about $20 million.

Wind is part of the problem during a blizzard. It piles snow in drifts and makes it difficult for people to navigate through city streets or over country roads.

BOWL OF SNOW

The greatest snowfall in a single storm in North America occurred at California's Mt. Shasta Ski Bowl from February 13 to 19, 1959. Over 15 feet of white stuff fell. That's over twice the height of basketball star Shaquille O'Neal.

SUPER-COOL ART ▶

People always rise to the occasion when nature sends mountains of flakes. If conditions are not disastrous, some people like to make snowmen or go sledding or skiing. Then there are the more serious creative types. Here, an artist in chilly Minnesota gets down to business sculpting a face of snow three stories high.

◀ AVALANCHE

An avalanche is an uncontrollable slide of ice and snow traveling hundreds of miles per hour. A volcano can cause an avalanche. The Mt. Saint Helens eruption on May 18, 1980, in Washington, caused 96 billion cubic feet of snow to go tumbling down.

▼ A snow-covered Mt. Saint Helens.

In 1947, traffic was at a standstill during the heaviest blizzard yet recorded in New York City.

▼ RECORD BREAKER

Unusual for New York City, streets were almost completely empty on January 8, 1996. The blizzard that hit caused the entire Northeast to come to a near stand-still. Schools were closed and businesses were shut tight.

In 1996, New York broke its 1947 record of 63.2 inches of total snowfall in one year with a chilling 75.6 inches.

FIRE!

It's true that people cause a lot of forest fires. Matches get thrown down on the dry forest floor, or a campfire is not extinguished properly. But nature sets its own fires, especially when the land is dry and the wind is blowing. In July 1994, in the western U.S., conditions were perfect for a fire. Lightning ignited trees, setting off a blaze that caused 240,000 acres to burn in 11 states. Fourteen firefighters died near Glenwood Springs, Colorado.

TALK ABOUT HOT!

It was 1871, the driest year in memory, and small fires were burning all over Wisconsin. Some got way out of control. In the forest fire that burned in Peshtigo, on October 8, 1871, about 1,200 lives were lost. Over 2 billion trees burned in what is considered the worst U.S. forest fire in history.

SMOKEY

His official name is Smokey the Bear. His official slogan is "Only *you* can prevent forest fires." In 1968, Smokey had become the most popular symbol in the U.S., beating out President Lyndon B. Johnson.

IN THE WOODS

In California, some homes are so close to the woods and forests that fires often overtake them. From October 20 to 23, 1991, a brush fire in Oakland destroyed over 3,000 homes and apartments, causing about $1.5 billion in damage.

LENDING A HAND

Over 10,000 fire-fighters from all over the United States came to help put out the 1988 fire in Yellowstone National Park. They cleared parts of the forest to create firebreaks, areas where there would be nothing the fire could feed on. But strong winds kept pushing the flames across the breaks.

Helicopters and airplanes often assist firefighters by dropping water and other fire retardants.

GOOD AND BAD

Some feel that forest fires may actually be a blessing. As the forest floor gets cluttered with underbrush and fallen trees, a potential fire becomes more and more dangerous. A fire tends to clean up the area and enables new trees to grow. And although wildlife flees the area during the time of danger, animals return once the fire is out.

▼ WALL OF FIRE

Yellowstone, one of the largest national parks in the United States, has over 2.2 million acres. In the summer of 1988, over 1.3 million acres burned—more than half the park! It was the worst forest fire in any of our national parks.

FIERCE FLOODS

Flooding in northern California reached the roofs in 1995.

Firemen are rescuing a man who went back to his flooded home in St. Louis to rescue his cat.

Rain is an essential part of our world. But when there's too much, we get floods and a number of other disastrous events. Rivers overflow and take houses and cars with them. Power lines go down, and fires get started. Then, mud may slide over the land.

During the flood in the Midwest in 1993, which left thousands of people homeless, farm animals and pets had to be rescued, too.

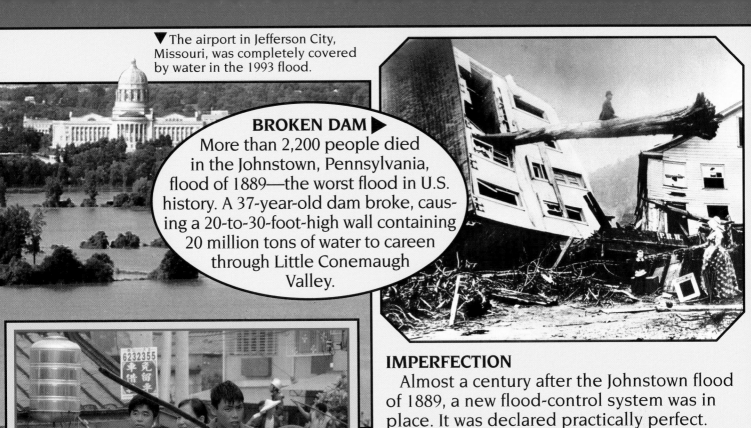

The airport in Jefferson City, Missouri, was completely covered by water in the 1993 flood.

BROKEN DAM ▶

More than 2,200 people died in the Johnstown, Pennsylvania, flood of 1889—the worst flood in U.S. history. A 37-year-old dam broke, causing a 20-to-30-foot-high wall containing 20 million tons of water to careen through Little Conemaugh Valley.

IMPERFECTION

Almost a century after the Johnstown flood of 1889, a new flood-control system was in place. It was declared practically perfect. Not so! Another flood swept over the land on July 19, 1977.

DANGEROUS RIVER

Southeast Asia experiences a lot of flooding. The Yellow River in China has overflowed many times. Major flooding there has caused the death of millions. In one flood alone, in 1931, 3.7 million people died.

IN A FLASH

Flash floods occur after heavy rains when the weather has been dry. This downpour causes lakes and reservoirs to overflow. Dry ground can't absorb the runoff fast enough.

In 1995, heavy rains flooded many European cities.

NATURALLY WACKY

When you think of natural disasters, you may think of earthquakes shaking down houses, floods covering the land, or tornadoes carrying off property. But some natural occurrences seem more wacky than disastrous.

A landslide buried this town in the Philippines after a volcanic eruption.

◄ Workers wade through a muddy volcanic aftermath in Colombia.

Heavy rains in California in 1995 created disastrous mudslides.

LANDSLIDE
Too much rain can turn soil to mud, and too much creates a landslide. Landslides can also be caused by earthquakes, which loosen rock and debris.

These kids have captured a few giant hailstones from a recent storm.

INCREDIBLE ICE BALLS
Sometimes perfect balls of ice, hail is one of those strange things in nature we're sometimes excited to see. But hail can be dangerous. In 1990, 60 people were injured by baseball-size hailstones in Colorado. It was a costly hailstorm, causing over $625 million in damage.

▼ REALLY BAD LUCK

Imagine being struck by lightning once. Disaster! Now imagine being struck by lightning seven times—and living to tell the tales. Former park ranger Roy C. Sullivan was struck in 1942 for the first time. Then, he was hit again in 1969, 1970, 1972, 1973, 1976, and 1977.

▼ One of many famous photographs taken of the dust bowl. Photographer Arthur Rothstein captured a father and his children running for cover as a storm of dust approached.

DUST BOWL

What happens when you don't have rain? Drought. The longest drought of the 20th century took place in the U.S. in the 1930s. During 1934, dry regions stretched from New York to California. Much of the Great Plains was called the "dust bowl." There, topsoil blew away easily because it had been over-worked. Strong winds carried it around and created drifts of fine dust.

▲WHALE OF A GALE

What is not a tornado or hurri-cane, but is not just a breeze? Strong winds known as gales. They shake whole trees and make it difficult for people to walk. These Japanese tourists got caught off guard in Paris by gusts of winds traveling over 80 mph!

WHAT A DEVIL!

Tornadoes suck up anything in their path. When they travel over the desert, sand is the thing they pick up most. This swirl of grit is called a dust devil, and its name suits it.

Photo Credits: Weird Science

R. Barlow: page 66
Dr. E.R. Degginger: page 69
Phil Degginger: page 68
Breck P. Kent: pages 63, 68
Gerald L. Kooyman: page 66
Carol Russo: page 59
David Scharf: pages 64-65
Dr. Kent A. Stevens/University of Oregon: page 69
John Lundberg, John Sullivan: page 66
AP Photo/Jean Clottes: page 70
AP Photo/Michael Stephens: page 63
AP Photo: page 73
UPI/Corbis-Bettman: page 63
EPConcepts/Custom Medical Stock: page 60
NIH/Custom Medical Stock: page 61
Larry Ulrich/DRK: page 61
Shahn Kermani/Gamma Liason: page 60
Xavier Rossi/Gamma Liason: page 77
Gorilla Foundation: page 67
The Granger Collection, New York: pages 58, 71
Michael Agliolo/International Stock: pages 62-63
Muybridge/Mary Evans Picture Library: page 59
Museum of Science, Boston: page 76
B. Balick (University of Washington) & NASA: page 72
D. Figer (UCLA) & NASA: page 73
International/NASA: page 73
Bates Littlehales/National Geographic Society: page 67
Joel Sartore/National Geographic Society: page 67
George Steinmetz/National Geographic Society: page 77
National Photo Service: page 71
The Natural History Museum/Orbis: page 69
Scott Camazine/Photo Researchers: pages 78
John Foster/Photo Researchers: pages 74-75
Tom McHugh/Photo Researchers: page 70
Will & Demi McIntyre/Photo Researchers: page 61
David Phillips/Photo Researchers: pages 58-59
Michael Covington/Phototake NYC: page 58
Roslin Institute/Phototake NYC: page 62
Peter A. Simon/Phototake NYC: page 79
Jean Claude Revy/Phototake NYC: page 63
NASA/Phototake NYC: page 74
Dr. Tony Brain/Science Photo Library: page 65
Chris Butler/Science Photo Library: page 72
Professors P.M. Motta & S. Correr/Science Photo Library: page 64
A.B. Dowsett/Science Photo Library: page 65
David Hardy/Science Photo Library: pages 72-73
James Holmes/Cellmark Diagnostics/Science Photo Library: page 62
James King/Science Photo Library: page 77
K.H. Kjeldsen/Science Photo Library: page 65
Mehau Kulyk/Science Photo Library: page 68
Matt Meadows/Science Photo Library: page 78
Astrid & Hanns Frieder Michler/Science Photo Library: page 60
Microfield Scientific LTD/Science Photo Library: page 60
National Library of Medicine/Science Photo Library: page 61
David Parker/Science Photo Library: page 64
John Reader/Science Photo Library: page 71
David Parker/Science Photo Library: page 79
Sandia National Laboratories/Science Photo Library: page 77
Science Photo Library: page 75
Dr. Steve Gull; Dr. John Fielden; Dr. Alan Smith/Science Photo Library: page 75
Dr. Linda Stannard, UCT/Science Photo Library: page 65
Geoff Tompkinson/Science Photo Library: page 79
Doe/Science Source: page 59
Kraft/Explorer Science Source: page 75
LLNL/Science Source: page 78
Alexander Tsiars/Science Source: page 76
David M. Phillips/Visuals Unlimited: page 64
Science VU/Visuals Unlimited: page 71

Scientific Consultant:
Malcolm Fenton, PhD.
The Dalton School

WEIRD SCIENCE

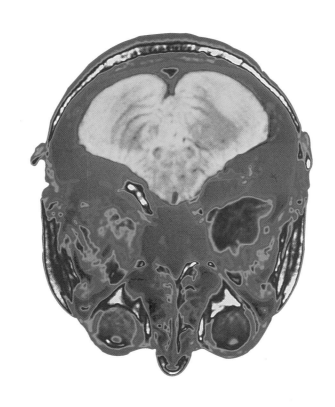

WHAT AN IDEA!

Once upon a time, it would have been as weird to believe that Earth is round as it would be now to believe it is flat. In science, curiosity and the need to solve problems lead to investigation. Then new information and technology can turn strange ideas into reality.

NEW VIEW

In 1609, Galileo looked at the moon through a new invention, the telescope, and saw craters. That may not be a shock to you. But at that time, people believed that Earth was unique, and at the center of the universe. To them, the moon was a heavenly globe of light. But Galileo discovered that the moon is a world that, in some ways, resembles our own. Galileo changed our ideas about the universe forever.

AN INSPIRING BATH ▼

In the 3rd century B.C., Archimedes had a difficult problem to solve—he needed to measure the volume of the king's crown, an irregularly shaped object. As this story is often told, Archimedes solved his problem while taking a bath. Climbing in, Archimedes realized that he was pushing out an amount of water equal to his body's volume. "Eureka!" he cried, meaning "I have it!" All he had to do was put the crown in water and measure the water it displaced.

A SMALL WORLD

Plagues have killed millions throughout history. But before the mid-19th century, no one knew about germs. When Louis Pasteur proved his germ theory, he convinced doctors to boil their instruments and wash their hands. One of medicine's greatest discoveries, it led to a doubling of life expectancy and a population explosion.

This microscope image (shown in background) is the bacteria E. *coli*, which lives naturally in the human intestine and is necessary to a person's good health.

HORSE SENSE ▲

In 1872, two men made a bet about whether all four of a horse's hooves leave the ground when it gallops. To settle the bet, photographer Eadweard Muybridge set up 24 tripwires and cameras along a racetrack. As the horse galloped, it triggered the cameras one after the other. One photo showed all four feet in the air.

CHANGING ▶ TRUTH

Democritus (460-370 B.C.) said that all matter was composed of particles so tiny that nothing smaller was conceivable. He called them atoms, meaning "indivisible." Too weird for the time, the idea wasn't taken seriously for 2,000 years. After all, you can't see atoms—250 million of them lined up measure only one inch! Even smaller are the subatomic particles that make up an atom—protons, neutrons, and electrons.

IT'S RELATIVE ▲

Albert Einstein changed our view of time. His theory said that if one twin traveled through space while the other stayed on Earth, both twins would feel time pass normally. But time would slow down for the traveling twin, and he would age less than his twin on Earth. An experiment with an atomic clock carried by a jetliner proved Einstein correct. The clock showed less time passing than a clock on Earth.

59

CURIOUS CURES

Until about 100 years ago, bloodletting was a normal treatment for many illnesses, including fevers and pneumonia. The doctor would cut the patient in a particular area of the body and let a certain amount of blood flow out. Now we think of this as weird, because loss of blood is harmful, not helpful. But some of the treatments we use today seem every bit as strange.

◀ IN THE DIRT

Where do you find wonder drugs? Scientists have poked around in the mud and muck to find them. The organism that produces streptomycin, an important antibiotic, was first discovered on a hen. Scientists then searched the whole farm and found the organism in the heavily manured soil around the henhouse.

OUCH! ▲

It may look painful, but acupuncture is a Chinese medical procedure that actually helps relieve pain. Even pets get relief. The physician may insert needles at more than 360 points in the body, and then twirl the needles or send electrical currents through them. In China, acupuncture is often used along with drugs during brain surgery.

◀ GROSS WORKS

Some ancient remedies are still used today. After reattaching fingers or toes, doctors may apply leeches—blood-sucking worms— to help keep tiny blood vessels from clogging. Maggots are used to remove dead flesh. And physicians sometimes pack deep wounds with the same thing used by Egyptians 4,000 years ago—sugar.

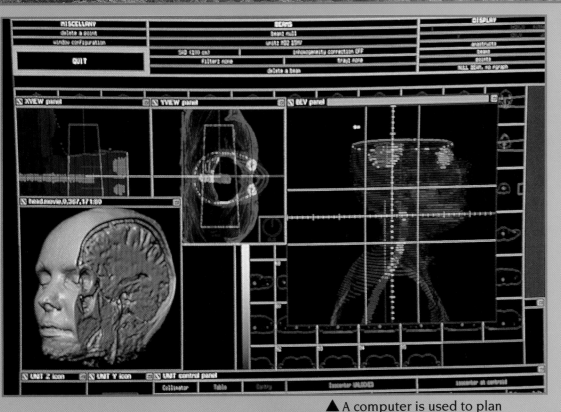

MISCELLANY
delete a point
window configuration
QUIT
BEAMS
beam null
units RBS 159V
SSD (100 cm)
filters none
tray none
delete a beam
DISPLAY
contours
beam
points
NULL SCAN. no signals
isohomogeneity correction OFF

XVIEW panel   YVIEW panel   DEV panel

head.movie.0.367.1/1:80

UNIT Z icon   UNIT Y icon   UNIT control panel
Collimator   Table   Gantry   Isocenter UNLOCKED   isocenter at controls

▲ A computer is used to plan radiation treatment.

KILL TO HEAL

Although X rays were first observed by accident, scientists have learned much about radiation and put it to good use. Too much radiation can be fatal to humans, but, by focusing those killing powers on diseased cells, radiation can be used to treat cancer.

GROWING CURES

We often think of plant remedies as something used only by ancient people, but almost 80 percent of people today rely mainly on this traditional medicine. Nearly 50 percent of medicines on the market come from plants. More than 25 percent of all medicines come from ingredients found in rainforest plants and animals.

◀ Marie Curie (1867-1934), the only person to win two Nobel Prizes in science, was the first to describe the process of radiation.

FROG MEDICINE

Long used by Ecuadorian Choco Indians as a weapon, the venom from the poison arrow frog can help as well as hurt. One compound in the venom acts as a painkiller that is 200 times better at fighting pain than morphine, a drug used in hospitals.

THE GENE SCENE

Our genes carry DNA (deoxyribonucleic acid)—the substance that makes us unique individuals. Although it fits into a single cell, the human DNA molecule is about six feet long. If you stretched out all the DNA molecules in a baby, and put them end to end, they would reach 114 billion miles—thirty times the distance between the Sun and Pluto!

SEEING DOUBLE ▲

In 1996, Dolly, a sheep in Scotland, became the first mammal ever cloned from adult cells. A clone is a genetic duplicate—an animal or plant that is identical to another. Scientists replace the DNA of an egg cell with the DNA from another cell. The egg then develops into a clone of the animal that provided the new genetic material.

◄ DNA DETECTIVES

Who did it? DNA can link criminals to a crime scene better than any fingerprint. Scientists use blood or skin cells found at the scene to create a genetic profile, which they compare to the suspect's. In one case, the DNA taken from cat hairs found at the scene of a crime was matched to that of the suspect's pet. The man was found guilty and convicted.

FAMILY TREES ▼

By using DNA, it's possible to trace family relationships. Scientists studying the DNA of a 9,000-year-old skeleton from Cheddar, England, discovered that he had a living descendant—the local school teacher! That's a family tree with long roots.

▲ ZONKEY!

What happens when you cross a zebra with a donkey? You get a crossbreed, an offspring with a mix of genes from both parents. Through biotechnology, scientists may isolate a specific gene from one species and give it to another, creating new varieties of plants and animals.

GENE ALARM

Do you wake up at the same time every day? We all have a biological clock that controls our sleep cycle. Researchers have discovered that, in mice, a gene controls this clock.

SAVING SPECIES

Test-tube tigers? Scientists are using frozen sperm and eggs to create baby animals that belong to endangered species. By fertilizing the egg in a test tube, it is possible to breed animals separated by great distances that might not otherwise mate.

▶ DINOS TODAY?

Bad news for *Jurassic Park* fans. In the movie, scientists used DNA taken from blood in the stomachs of biting insects trapped in amber (fossilized tree resin) to re-create dinosaurs. In reality, this would be impossible. However, scientists have discovered some pieces of dinosaur DNA in fossil bones.

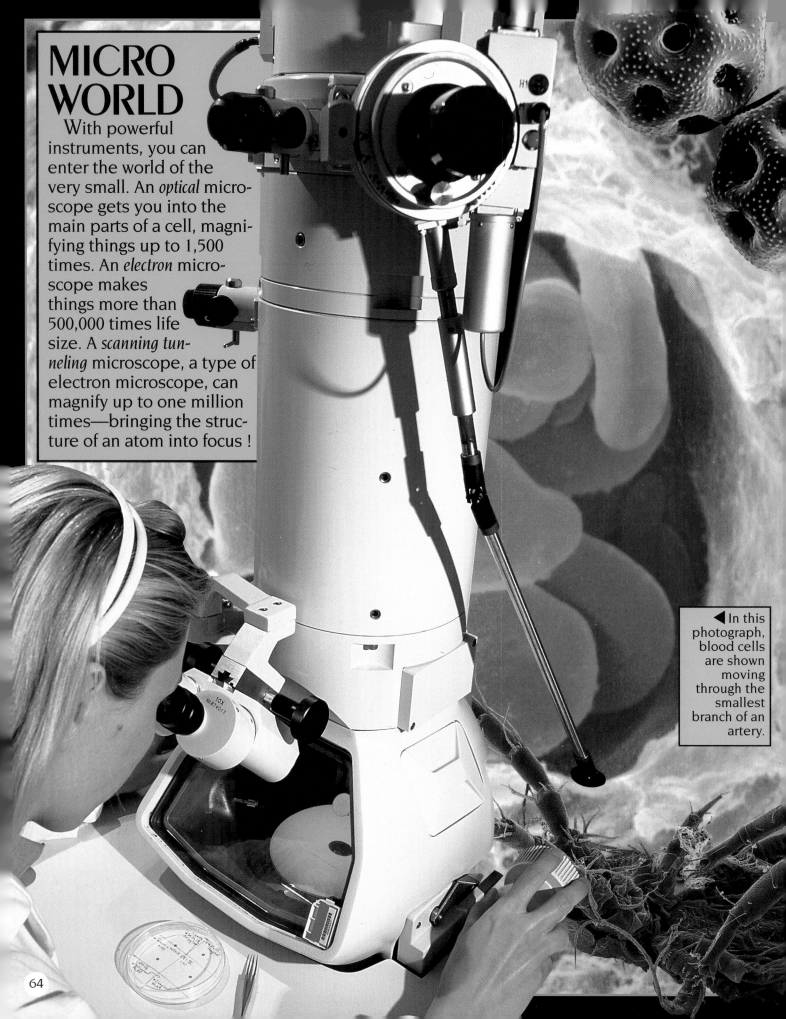

MICRO WORLD

With powerful instruments, you can enter the world of the very small. An *optical* microscope gets you into the main parts of a cell, magnifying things up to 1,500 times. An *electron* microscope makes things more than 500,000 times life size. A *scanning tunneling* microscope, a type of electron microscope, can magnify up to one million times—bringing the structure of an atom into focus!

◀ In this photograph, blood cells are shown moving through the smallest branch of an artery.

▲Pollen can cause allergic reactions known as hayfever.

AH-CHOO!

The body has a remarkable defense against infection. But in some people, this immune system gets a little confused. It recognizes certain harmless particles as dangerous and tries to fight them off. An allergic reaction is the result, which can be minor, like a sneeze, or, in extreme cases, life-threatening.

MICRO WARFARE

Viruses represent one of the biggest challenges to overcoming infectious diseases. When a virus, such as the common cold, invades a cell, it forces the cell to make copies of the virus. To attack the virus is to attack the body's own living cells. No cures exist, only preventive vaccines.

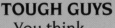

▲This weird-looking virus is responsible for causing warts.

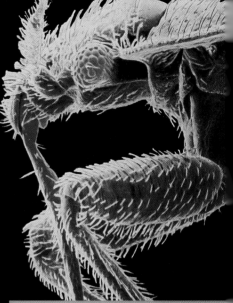

TOUGH GUYS

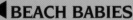

You think you're tough? Some bacteria live in boiling hot water. Others survive radiation that would kill a human. Scientists have even found bacteria living 4,500 feet below the ground. These bacteria appear to survive in total darkness on nutrients they extract from the rocks.

MITE-Y SMALL

You never see them, but the tiny spider relatives known as mites (left) are all around us, even in the cleanest house. They crawl through carpet, prowl your bed, and even live in your hair! But don't confuse them with the blood-sucking, foul-smelling insects known as bedbugs (above). Mites actually help keep your house clean by feeding on flakes of dead skin.

◀BEACH BABIES

It's just a beach, or is it? Living in the sand is one of the richest animal communities found on Earth. These tiny meiofauna (MY-oh-faw-nah) are strange creatures—some have heads covered with whirling hairs, others cling to sand grains with hooked claws. A single handful of wet sand may hold 10,000 of these animals.

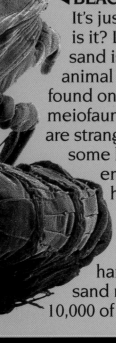

NATURAL ODDITIES

Imagine finding a new life form, or talking to animals. In nature, scientists encounter some pretty strange things, and engineer some weird ways of getting inside an animal's world.

▲DOWN DEEP

In the murky waters of the Amazon River, where light penetrates to only a few inches below the surface, many new species of fish are being discovered. Mostly blind, they rely on electrical signals to navigate the dark river water.

WHITE WONDER▼

Coloration exists for a reason in nature, often as camouflage against predators. That's why it's strange to find an animal colored differently than the norm—like this pure white emperor penguin spotted in Antarctica. Sometimes white animals are albinos, which lack pigment and have red eyes. However, this penguin is not an albino, but a kind of genetic mutant rarely seen.

◀ SPYING CRABS

How do you know if an animal is using its eyes, and not its sense of smell? Some scientists mounted a tiny spy camera above the eyes of a male horseshoe crab to make a film of what the crab saw—female crabs. Meanwhile, they monitored electrical responses in the nerves that connect the eyes to the brain. Back in the lab, male crabs who saw the film responded in much the same way as the crab on the loose, proving that the eyes were at work.

PUPPET MOM ▶

Raising animals in captivity leads to all sorts of trouble. The animals see humans as protective parents and never learn to survive on their own. One solution to raising bald eagles and other birds is to use puppets that look like the adult animals. The babies receive food from a beak, not a human hand.

SMART TALK ▲

Have you ever wondered if animals have their own language? A study on prairie dogs used a computer to correlate squeaks and chirps with events happening at the time. Results suggested that prairie-dog talk distinguishes a coyote from a German shepherd and a man from a woman.

PLANT FIGHT

When disease strikes, plants fight back. At least, that's what some scientists think. Plants produce the chemical salicylic acid, or aspirin, when they are sick. This self-made medicine boosts a plant's immune system. It also stimulates other plants to be on the defense. In the future, instead of spraying toxic pesticides, maybe people will treat plants with aspirin.

▼ARTY APES

Believe it or not, people can communicate with animals. The famous gorilla named Koko began learning sign language in the 1970s. She uses more than 1,000 signs to relay her thoughts. One of her favorite activities is painting. When asked what her blue, red, and yellow painting was, Koko answered, "bird."

LOST IN TIME

When the first fossils were found, no one knew what they were. Fossils are traces of plants and animals preserved in the earth. Today, scientists called paleontologists (pay-lee-on-TALL-oh-jists) are unraveling the stories that fossils tell.

STRANGE CREATURES

Preserved in 530-million-year-old rock in Canada, called the Burgess Shale, are some of the weirdest animals that ever lived on Earth. One was a worm with five eyes and a trunk.

FEATHERED FINDS ▲

A rich fossil bed discovered in China in the 1990s is remarkable because of the amount of detail preserved in the rock. Among its treasures are lots of bird fossils showing feathers, including the oldest beaked bird.

HUGE INHABITANTS ▼

Fossils reveal that South America had some hefty inhabitants—an eight-ton, meat-eating dinosaur we call Giganotosaurus, and the plant-eater Argentinosaurus (below), which weighed in at around 100 tons. Other discoveries include two-foot spiders and a forty-foot-long crocodile.

ANCIENT FOREST

A subtropical forest only 680 miles from the North Pole? On Axel Heiberg Island, scientists found the *mummified* remains of one. The forest grew 45 million years ago when the area was warm and swampy. The really weird thing is that the wood could still burn today.

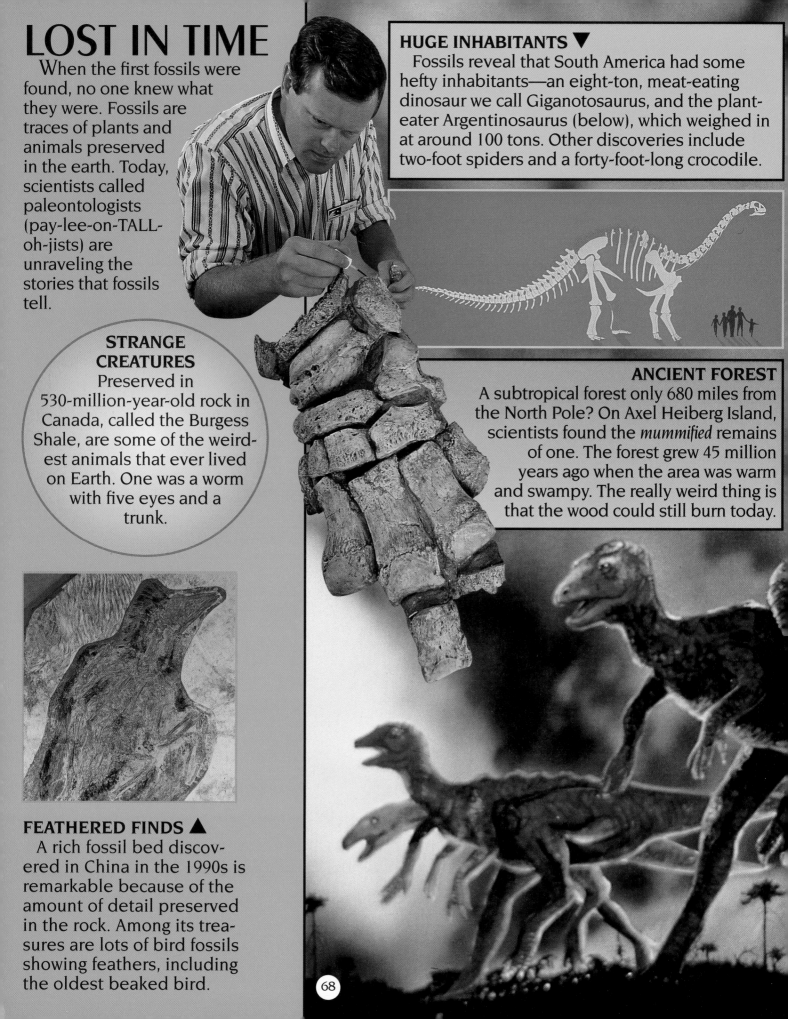

UP IN THE AIR ▶

If you had been in Texas about 65 million years ago, you might have seen something impressive overhead—the largest flying creature ever, a reptile called Quetzalcoatlus (KWET-zal-KWA-tel-us). It weighed about 190 pounds and had a wingspan of about 40 feet—the size of a small plane.

▼ CYBER DINOS

If you want to see how dinosaurs moved, you have to bring them to life. Some scientists have done just that—in cyber space. They used software to reconstruct sauropods, such as Apatosaurus (also known as Brontosaurus). This 100-ton dino may have broken the sound barrier, creating a sonic boom by using its 3,500-pound tail like a whip!

THE GREAT DYING

The dinosaurs died out 65 million years ago, but about 245 million years ago, a much larger extinction event took place. During what is known as the Great Dying, as much as 96 percent of all plant and animal species on Earth may have been wiped out.

COLORFUL DINO?

Ever wonder what color Stegosaurus really was? Researchers looking at fossils have identified structures and pigments responsible for colors. Using this information, they've determined that a 370-million-year-old fish was dark red on top and silver underneath. Could the dinosaurs be next?

DIGGING HISTORY

Scientists are still trying to piece together human history, especially the prehistoric period when no written language existed and few records were kept. Archaeologists and anthropologists are the detectives looking for the clues.

OLD ART ▲

The last time human eyes saw these cave paintings, discovered in 1994 in the Ardeche River canyon of France, may have been 30,000 years ago. Over 300 paintings represent Ice Age animals that include horses, bears, rhinoceroses, and bison.

▼ MODEL PAST

Starting with a skull and using charts that show the thickness of tissue, scientists can re-create faces from long ago. They build models of the face from clay or use computers to make a three-dimensional image.

JIGSAW HEAD ▲

Bones and artifacts, discovered on a dig or purely by chance, can take scientists back millions of years. Richard Leakey and his team of researchers found 150 skull fragments and other bones in Kenya dating almost 2 million years ago. The team reconstructed the skull, and named it "handy man" because it was found alongside stone tools.

◀ PEKING MAN

A famous collection of fossils known as Peking Man was discovered in 1921 in China. Thought to be 400,000 years old, the bones were the first of their kind found in Asia. Stone tools were also found in the cave.

VIKING ◀ VISITS

Did some European explorer beat Columbus to the Americas? The remains of eight buildings and other artifacts discovered in Newfoundland, Canada, in 1960 finally confirmed that Vikings got here long before 1492. But their attempt to settle the continent failed.

ICE MAN

Imagine hiking in the Alps and finding a body that's over 5,000 years old, preserved like a time capsule in a glacier. That's what happened in 1991, when two mountaineers discovered the "Ice Man."

BOG BODIES

In cool places where water stands still, layers of dead plants pile up sometimes 40 feet thick underneath living plants, and a bog is formed. Called peat, the dead plant matter is useful as a kind of fuel. People digging for peat have made some bizarre discoveries. They have found perfectly preserved bodies—whiskers and clothes included—dating from over 2,000 years ago.

HANDLE WITH CARE

Scientists can use high-tech methods to learn all about a mummy without damaging it. Using Computer Assisted Tomography (CAT) scanning, they take a series of X-ray pictures of the mummy still in its wooden case. Then the computer adds up the images to create a 3-D image.

SPOOKY SPACE

In the vast reaches of space, almost everything, like the death of a star, seems weird and mysterious. At the end of its life, a star may collapse in upon itself and form a black hole. Sounds interesting, but you don't want to visit. A black hole's gravity is so strong that it pulls in everything around it, even light.

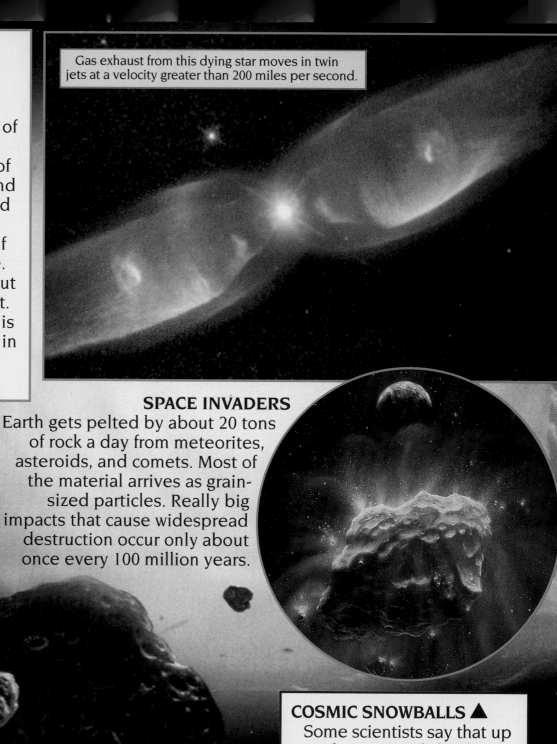

Gas exhaust from this dying star moves in twin jets at a velocity greater than 200 miles per second.

SPACE INVADERS

Earth gets pelted by about 20 tons of rock a day from meteorites, asteroids, and comets. Most of the material arrives as grain-sized particles. Really big impacts that cause widespread destruction occur only about once every 100 million years.

COSMIC SNOWBALLS ▲

Some scientists say that up to 30 house-sized chunks of ice are hitting our planet's atmosphere every minute. Each of these big snowballs weighs about 20 tons. We're not up to our noses in snow because the ice melts while still thousands of miles above Earth.

▶BIG LIGHT

The Pistol Star cannot be seen without a telescope, yet it sends out 10 million times more energy than our Sun! Newly discovered, it is believed to be 186 million to 280 million miles across—about as big as Earth's orbit around the Sun.

ANOTHER COMPANION

Earth travels with more than just a moon—Asteroid 3753, just over three miles in diameter, is trapped by Earth's gravity into a strange orbit between Mercury and Mars. It takes 770 years for this asteroid to orbit Earth.

◀HOT BLOB

When electrically charged gas gets ejected from the Sun, a blob of it may weigh tens of billions of tons, travel about 2.2 million miles per hour, and carry enough energy to boil off the Mediterranean Sea. When it hits Earth's atmosphere, a geomagnetic storm occurs, sending curtains of colorful light called auroras over the polar skies.

◀HOME UNKNOWN

What is the best-mapped planet in our solar system? Not Earth! More than 98 percent of Venus's surface has been mapped, thanks to the explorer spacecraft *Magellan*. Here on Earth, more than two-thirds of the surface is covered with water, and much of it cannot be charted as accurately as the Venusian surface.

EERIE EARTH

Our home planet harbors a store of weird happenings, from powerful earthquakes to mysterious, phenomenal events. In 1811, things were rocking in New Madrid, Missouri. The area experienced the strongest earthquakes in recorded American history. More than 3,000 square miles of land were visibly damaged and, for a brief time, the Mississippi River flowed backwards!

SPIN CITY

The Moon's gravitation acts like a brake on Earth, slowing down its rotation. Each year our days get about 20 millionths of a second longer. But something else is having the reverse effect. Scientists have discovered that dams, in concentrating water away from the equator, are making the planet spin faster!

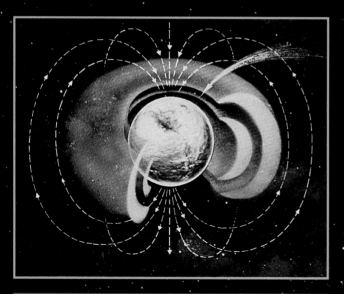

COMPASS CHECK ▲

The motion of molten metal in Earth's outer core produces an electric current, causing the planet to behave like a giant magnet. Earth's magnetic field extends about 37,000 miles into space and protects us from some of the Sun's most harmful particles. Curiously, the magnetic field reverses from time to time—the last occasion was about 700,000 years ago.

HIDDEN LAKE

A freshwater lake about the size of the state of New Jersey lies under more than two miles of ice in eastern Antarctica. The lake may be a mile deep in places. Bacteria and other life forms in Lake Vostok haven't been in contact with the surface for at least 50,000 years, and perhaps as long as three million years.

200 Million Years Ago

100 Million Years Ago

Today

▲ON THE MOVE

Think you can stand perfectly still? Guess again. Not only is Earth rocketing around the Sun at 66,600 miles per hour; it's spinning on its axis at about 1,000 miles per hour. Even the ground beneath you is never motionless, as big slabs of Earth's crust, called tectonic plates, are moving very slowly.

◄ Scientists believe that the continents were once all joined together, and that the movement of Earth's crust caused them to break apart. The Americas continue to drift away from Europe and Africa at a rate of one inch per year.

BIG BANG

When the Indonesian volcanic island of Krakatau blew itself to bits in 1883, a sound like distant cannons reached places nearly 3,000 miles away! The blast caused huge waves, called tsunamis (sue-NAH-mees), as far away as South America and shot ash an estimated 50 miles into the air.

PINGO PONG

In Siberia and northern Canada, frost stays in the ground year round, reaching as deep as 1,650 feet. Houses have to be built on stilts so they don't melt the frost and sink into the mud. Sometimes the frost pushes up piles of earth as high as 230 feet to create mounds called pingos.

DREAMS TO DEVICES

Scientists have built some weird and amazing things. Researchers in California have created the most powerful magnet in the world, 250,000 times as strong as Earth's magnetic field. Other scientists have created a light brighter than every star in our galaxy, concentrated in a spot the size of a pinhead. Called the Vulcan laser, it may enable scientists to look into living cells and capture molecules in action.

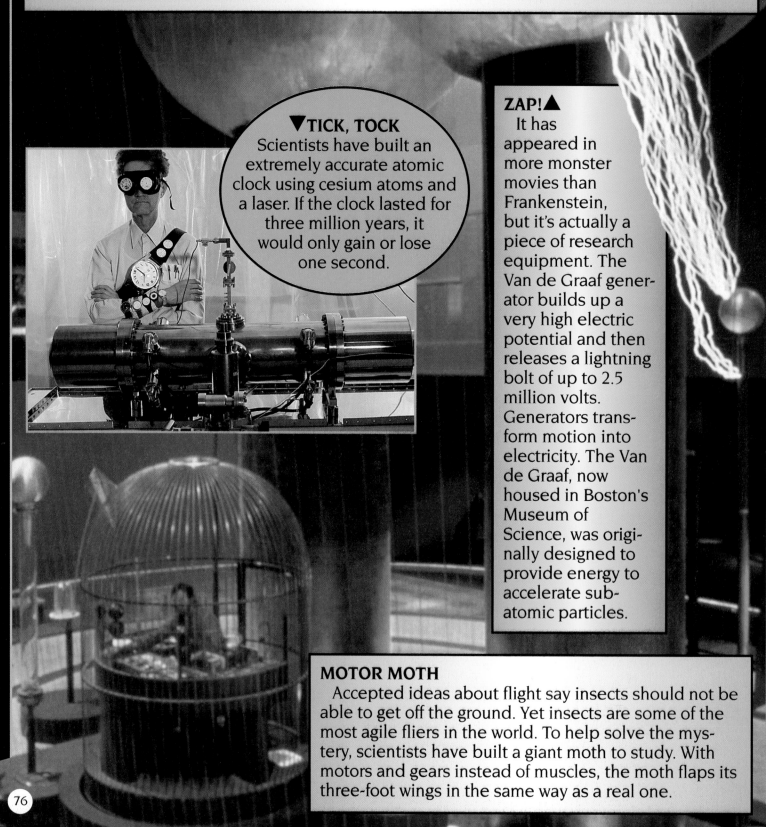

▼TICK, TOCK
Scientists have built an extremely accurate atomic clock using cesium atoms and a laser. If the clock lasted for three million years, it would only gain or lose one second.

ZAP!▲
It has appeared in more monster movies than Frankenstein, but it's actually a piece of research equipment. The Van de Graaf generator builds up a very high electric potential and then releases a lightning bolt of up to 2.5 million volts. Generators transform motion into electricity. The Van de Graaf, now housed in Boston's Museum of Science, was originally designed to provide energy to accelerate sub-atomic particles.

MOTOR MOTH
Accepted ideas about flight say insects should not be able to get off the ground. Yet insects are some of the most agile fliers in the world. To help solve the mystery, scientists have built a giant moth to study. With motors and gears instead of muscles, the moth flaps its three-foot wings in the same way as a real one.

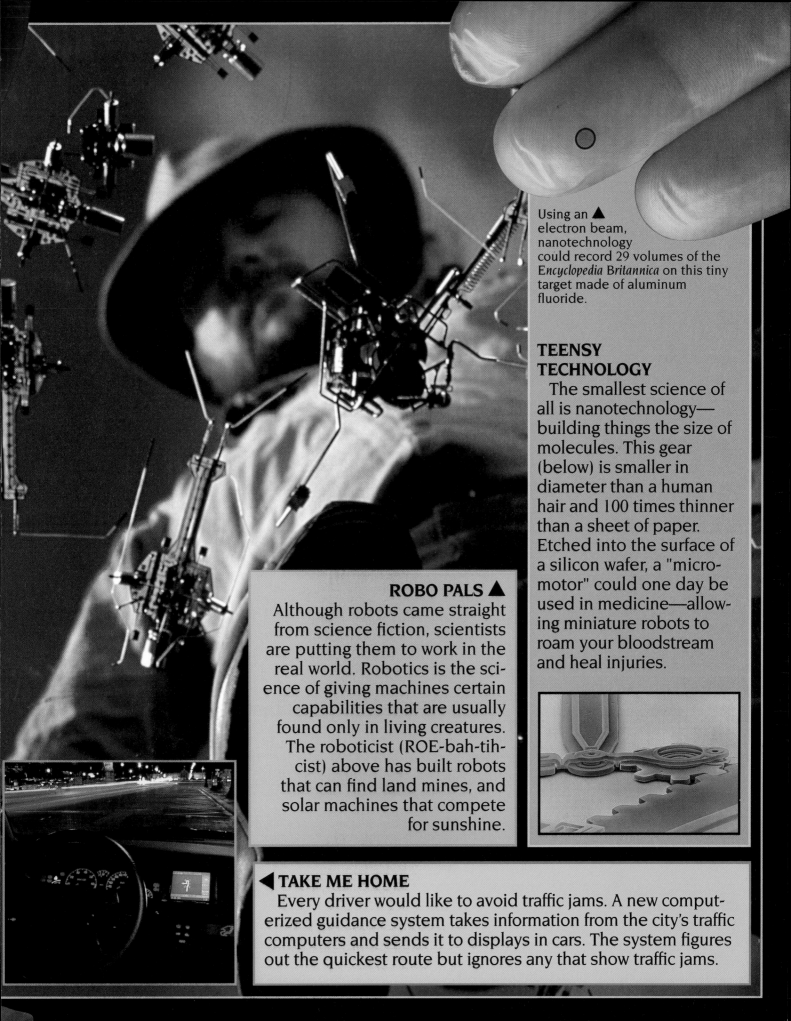

Using an ▲ electron beam, nanotechnology could record 29 volumes of the *Encyclopedia Britannica* on this tiny target made of aluminum fluoride.

TEENSY TECHNOLOGY

The smallest science of all is nanotechnology—building things the size of molecules. This gear (below) is smaller in diameter than a human hair and 100 times thinner than a sheet of paper. Etched into the surface of a silicon wafer, a "micromotor" could one day be used in medicine—allowing miniature robots to roam your bloodstream and heal injuries.

ROBO PALS ▲

Although robots came straight from science fiction, scientists are putting them to work in the real world. Robotics is the science of giving machines certain capabilities that are usually found only in living creatures. The roboticist (ROE-bah-tih-cist) above has built robots that can find land mines, and solar machines that compete for sunshine.

◀ TAKE ME HOME

Every driver would like to avoid traffic jams. A new computerized guidance system takes information from the city's traffic computers and sends it to displays in cars. The system figures out the quickest route but ignores any that show traffic jams.

STRANGE STUFF

Sometimes when we think of science, images of smoking beakers come to mind. In the laboratory, scientists can discover or create some pretty weird things.

▼ Solid carbon dioxide, known as dry ice, becomes a gas when mixed with water, producing thick clouds of fog.

◀ **BUCKYBALLS?**
Sixty carbon atoms arranged in a perfect sphere make up a molecule of buckminsterfullerene, or a "buckyball." Because of their shape, these molecules have been named after Buckminster Fuller, an engineer who developed the domed stadium. Buckyballs can be found in the soot floating around after you blow out a candle. Other such hollow carbon molecules are called bunny balls, bucky-babies, and bucky onions.

WHAT'S THE MATTER?
Solid, liquid, gas, plasma—the four states of matter. Now there's a weird fifth form—clusters. Clusters are groups of atoms that seem to be somewhere between atoms and the regular-sized world. They have strange electrical, optical, and magnetic properties.

STRETCHED THICK
If you punch a pillow, your fist leaves a dent in the surface. But some strange substances called auxetic (og-ZED-ik) materials don't act this way. When stretched, they get thicker. A pillow made of auxetic material would expand when punched.

◀ **LIGHTWEIGHT**
Scientists have made silica, the raw material for glass, into something new—an aerogel (AIR-oh-jell). The substance contains as much as 96 percent air and weighs far less than glass. Because it's so clear, aerogel is hard to find once it is placed on a lab bench. Here, it is shown supporting a weight and a penny.

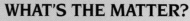

GENIE IN A BOTTLE?

No, it's not magic. It's a hologram, a photograph taken with laser light. One set of light waves is sent to the object, which reflects the light onto film. Another set of light waves, sent directly to the film, intersects the reflected light waves and creates a 3-D image. You can walk all the way around the picture to view it from different angles.

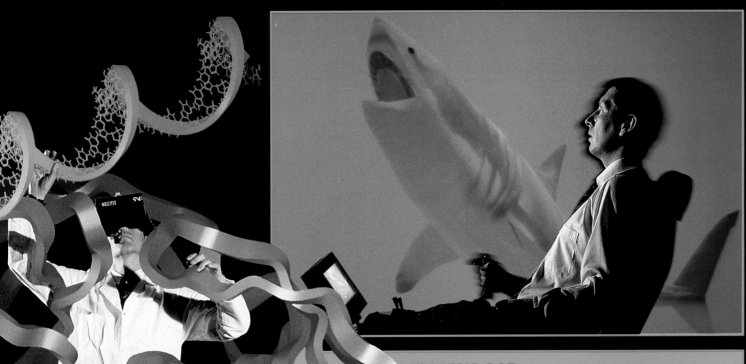

VIRTUAL LAB

Sometimes it's difficult, or completely impossible, for scientists to conduct their studies in real-life places. Virtual reality, simulated by computers, takes them there—or appears to.

◄ In a virtual molecular interaction, a researcher can pick up molecules, reorient them, and study what occurs.

79

Scientific Consultant:
Amy Gallagher
Hayden Planetarium
American Museum of Natural History

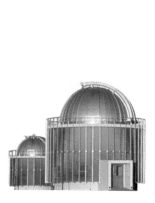

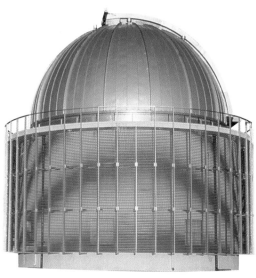

SPACE

INTO SPACE

Our world seems very large, but Earth represents an incredibly small place in the universe. Outside our planet is the rest of our solar system, which is contained within the Milky Way galaxy. If you could travel at the speed of light, 186,000 miles per second, it would take you 100,000 years to cross the Milky Way. And beyond it is the rest of the universe—billions and billions of other galaxies.

SHAPELY GALAXIES ▲
Galaxies have shape. The Milky Way is spiral shaped, like a whirlpool. Some galaxies are *elliptical*, or oval shaped. In this photo, many galaxy shapes and colors can be seen. The photo, taken by the Hubble Telescope, is the deepest view we've ever had of the universe.

▼A spiral galaxy.

THE WANDERERS
Planets are always moving, revolving around the Sun. The ancient Greeks named these restless objects *planets*, which means wanderers. Today we know of nine in our solar system—Mercury, Venus, Earth, Mars, Jupiter, Saturn, Uranus, Neptune, and Pluto. Moons also "wander," or revolve, around their planet.

SOLAR CENTER

For a long time, people believed the Earth to be at the center of all heavenly bodies. Then, in the fifteenth century, a Polish monk named Copernicus suggested that the Sun was at the center. People were shocked by this theory.

SKY LIGHTS

For your first space adventure, look into the night sky. Even without the aid of a telescope, you can see our Moon, the five planets closest to us, and more than 5,000 stars. You can even see the glow from another galaxy, called Andromeda.

GRAVITY GLUE

Ever wonder what holds our solar system together? Isaac Newton figured out the answer. An attraction, or pull, which he called *gravity* keeps the planets circling, or *orbiting*, the Sun. If it weren't for the gravity between the Sun and planets, the planets might move through space in a straight line. Also, Earth's gravity is what holds you to the ground.

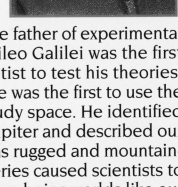

▲ This color-enhanced picture of the Moon was taken by the *Galileo* spacecraft, which was named for the great astronomer. Blue and orange shades indicate volcanic lava flows.

THE GREAT GALILEO

Known as the father of experimental science, Galileo Galilei was the first scientist to test his theories. In astronomy, he was the first to use the telescope to study space. He identified four moons of Jupiter and described our Moon's surface as rugged and mountainous. His discoveries caused scientists to think of planets as being worlds like our own Earth.

MINOR PLANETS

There are also jagged and round chunks of rock known as *asteroids* revolving around the Sun. These are minor planets, and there are thousands of them located between the orbits of Mars and Jupiter in an asteroid "belt." But some are outside the belt and sometimes pass very near to Earth.

83

UP, UP, AND AWAY!

It's only human to want to explore other worlds. Scientists started out viewing space from the ground, and telescopes brought them a little closer to other planets. But some people dreamed of actually going there. That's how space travel began.

A science-fiction book that really inspired early rocket scientists was Jules Verne's *From the Earth to the Moon*, published in the 19th century.

STEP ROCKETS ▲

In 1926, Robert H. Goddard became the first to build and launch a liquid-fuel rocket, which traveled 184 feet. He also created the "step" rocket. As fuel was used up in one stage, that section of the rocket would be dropped. It was one of the most significant inventions for getting rockets beyond the gravity of Earth.

◀ SHORT FLIGHT

The U.S. had trouble getting a *satellite* into space. Satellites are objects or vehicles launched into space to orbit the Earth. One called the *Vanguard* was ignited for launch in 1957, but the rocket rose only four feet, fell back to the launch pad, and exploded.

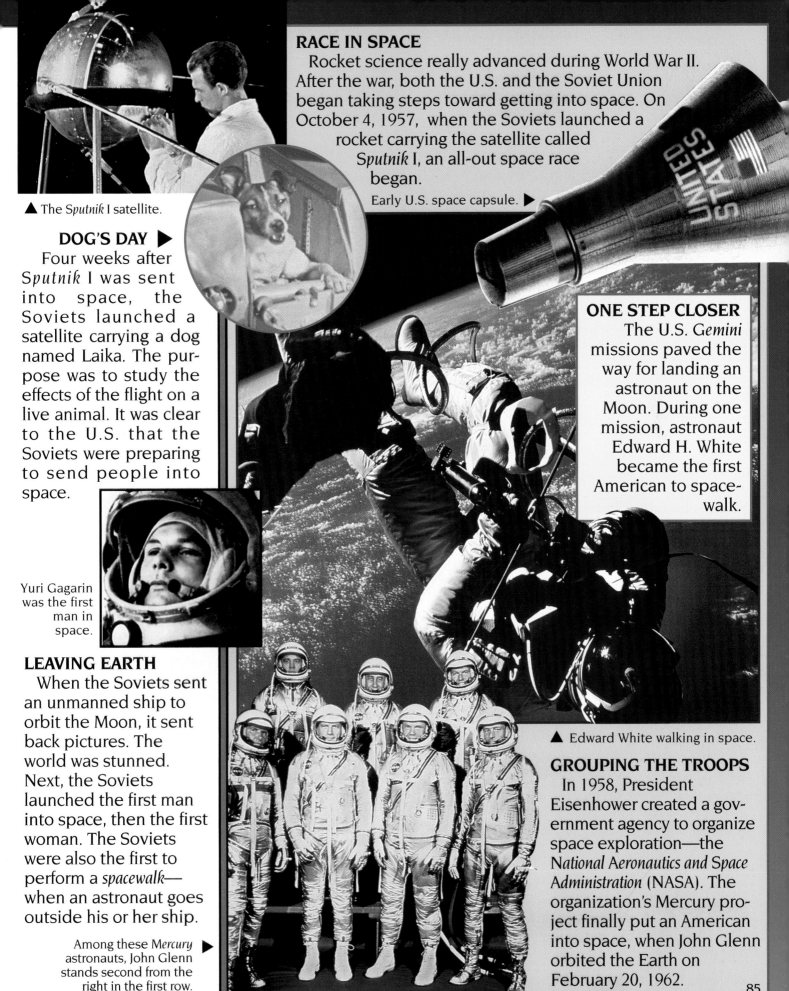

▲ The *Sputnik* I satellite.

RACE IN SPACE

Rocket science really advanced during World War II. After the war, both the U.S. and the Soviet Union began taking steps toward getting into space. On October 4, 1957, when the Soviets launched a rocket carrying the satellite called *Sputnik* I, an all-out space race began.

Early U.S. space capsule. ▶

DOG'S DAY ▶

Four weeks after *Sputnik* I was sent into space, the Soviets launched a satellite carrying a dog named Laika. The purpose was to study the effects of the flight on a live animal. It was clear to the U.S. that the Soviets were preparing to send people into space.

Yuri Gagarin was the first man in space.

ONE STEP CLOSER

The U.S. *Gemini* missions paved the way for landing an astronaut on the Moon. During one mission, astronaut Edward H. White became the first American to space-walk.

LEAVING EARTH

When the Soviets sent an unmanned ship to orbit the Moon, it sent back pictures. The world was stunned. Next, the Soviets launched the first man into space, then the first woman. The Soviets were also the first to perform a *spacewalk*— when an astronaut goes outside his or her ship.

Among these *Mercury* ▶ astronauts, John Glenn stands second from the right in the first row.

▲ Edward White walking in space.

GROUPING THE TROOPS

In 1958, President Eisenhower created a government agency to organize space exploration—the *National Aeronautics and Space Administration* (NASA). The organization's Mercury project finally put an American into space, when John Glenn orbited the Earth on February 20, 1962.

85

MAN ON THE MOON

Up until the 1960s, the Moon was an unknown world, and people had seen it only at a great distance, about 240,000 miles away. Then the United States' Apollo 11 mission landed there. The whole world got a close-up picture of the gigantic craters and rocky surface of the lunar landscape.

ON THE DARK SIDE OF THE MOON

The lunar module called the *Eagle* landed on the dark side of the Moon on July 16, 1969. Neil Armstrong emerged, stepped down a ladder to the Moon's surface, and said, "That's one small step for [a] man, one giant leap for mankind."

▲ The *Apollo* 11 blastoff.

EASY DOES IT

Walking on the Moon is fairly easy, but it's bouncy. Gravity is one-sixth of what it is on Earth, and things seem to weigh that much less. If you weigh 100 pounds on Earth, you'd weigh about 16 pounds on the Moon. The *Apollo* astronauts felt light and bobbed along as they explored.

◄ Scientists at NASA had feared that astronauts would have to wade through several feet of Moon dust just to get around. But footprints proved there was only a fraction of an inch of this powdery soil.

SPLASHDOWN

Once the Moon work was complete, the *Eagle* met up with the command module *Columbia*. For the next 60 hours, the spaceship traveled home. It entered Earth's atmosphere in a shower of fire at 25,000 miles per hour, before splashing down in the Pacific.

Apollo 11 astronauts were given a ticker-tape parade in New York City to welcome them home.

MISSION CONTROL ▼

Mission controllers on the ground communicate with astronauts millions of miles away from Earth.

MOON BUGGY

Just two years after the first landing, a car was brought along to the Moon, the *Apollo 15 Rover*. This battery-powered buggy had tires made of piano wires. Although practically weightless—about 76 pounds on the Moon—it really dug into the lunar dust.

LUNAR LABOR

Part of the astronauts' mission was to collect rock and soil for study back home. With no organisms having ever been found, the Moon is said to be a "dead planet." But it's an *old*, dead planet. Rock collected by Apollo 15 astronauts proved to be about four billion years old!

NEW DIRECTIONS

The success of the Apollo missions encouraged NASA to make more plans for space exploration. Among them were plans to build a space station and a shuttle. Another goal was to send unmanned probes to explore other planets.

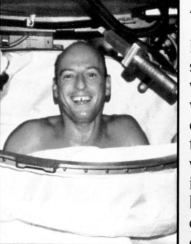

▲ Astronaut Sally K. Ride, the first U.S. woman in space, made a flight on the *Challenger* in 1983.

▲ *Skylab* offered the first shower in space.

▲ EYES IN THE SKY

After the Moon landing, NASA launched *Skylab*. This station floating in space would allow greater observations of the Sun, which can't be viewed directly with the human eye. Also, it would provide clearer images of stars. Out in space beyond the wavering gases of Earth's atmosphere, stars don't appear to twinkle.

FIRST FLIGHT

The shuttle was a masterpiece of engineering. After being launched into space by a rocket, it could be flown and landed like a plane. In 1981, *Columbia* was the first shuttle to make a space flight.

88

UNMANNED PROBES

NASA has launched many unmanned ships to explore space. These probes are less expensive. They also expand our knowledge of space without endangering human life. Between 1969 and 1978, 13 *Pioneer*, *Mariner*, *Viking* and *Voyager* probes were shot into space. In 1996, the *Galileo* probe landed on Jupiter.

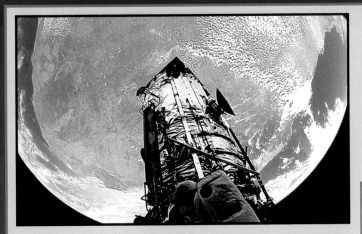

◄ PAYLOAD

The shuttle was designed to transport equipment as well as astronauts into space. Its total payload could weigh up to 65,000 pounds. In 1990, it carried and released the Hubble Space Telescope. This orbiting telescope provides clear pictures of stars, planets, and other heavenly bodies.

REPAIR CREW ►

One very important job done by shuttle crews is repair work. A robotic arm on the shuttle helps bring satellites into the cargo bay. The first satellite repair work was done in 1984 to the *Solar Max*. George D. "Pinky" Nelson and James D. "Ox" Van Hoften repaired the astronomy satellite and released it back into orbit.

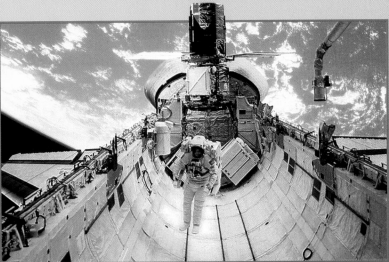

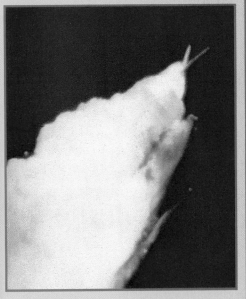

HUMAN SATELLITE

With one new high-tech piece of equipment, a couple of astronauts became the first human satellites. In 1984, the *Manned Maneuvering Unit* (MMU) allowed astronauts Bruce McCandless and Robert Stewart to do a spacewalk without being attached to the ship. McCandless flew 320 feet from the *Challenger*.

▲ MISSION OF DISASTER

The space shuttle was incredibly reliable for 24 flights. Then, on January 28, 1986, the *Challenger* exploded ninety seconds into its flight, killing seven crew members. The tragedy was heart-stopping, and reminded the public how dangerous space exploration can sometimes be.

LOOKING TOWARD HOME

At the time of the first Moon landing, space was considered the final frontier—the place where new worlds would be discovered. No one really imagined the impact space technology would have back on Earth. Today, it's hard to imagine living *without* satellites.

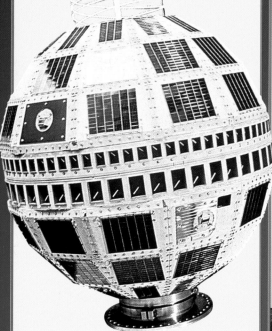

In 1962, *Telstar* I became the first active communications satellite. It also transmitted the first live television broadcast across the Atlantic.

EASY LISTENING
Satellites have made communication between continents easier and faster. In 1956, voice cable laid across the bottom of the ocean could only handle 36 phone calls at a time. Today, the most advanced international satellites can carry more than 120,000 calls at a time.

Some satellites remain stationary in space, while others orbit the Earth.

FIRE DOWN BELOW
Earth-survey satellites watch the environment. They measure quantities of water and ice, monitor coastal water pollution, and evaluate soil. When there's a fire in the forest, these satellites will detect it.

DISH CITY
Today, more than 3 million Americans have satellite dishes that pick up television transmissions. Dishes are also used by broadcast stations, which transmit information to satellites.

When space photos showed a royal blue sphere suspended in space like a jewel, Earthlings got a new view of their own planet.

HURRICANE HUGO
31.4N 78.4W
21 SEPT 135 MPH

KILLER STORM ▲

On the news at night, weather broadcasters often show pictures of cloud patterns moving across the Earth. These satellite photos help predict rain, snow, and severe storms. Hurricane Hugo was one of the most powerful storms of the century. But it killed only 40 people, because satellites gave residents an early warning.

1987

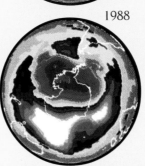

1988

◄ GLOBAL WARMING

Earth's atmosphere is composed of five layers of gases that surround the planet. They block much of the Sun's radiation. However, holes have developed in the ozone layer because of Earth's pollution. The two satellite images at left compare the size of the holes in 1987 and 1988.

SPACE JUNK

Useful machine satellites are not the only things orbiting our planet. There's also junk up there—spent rocket boosters, snippets of wire, and even paint—left behind by decades of space missions. Traveling 25,000 miles per hour, the debris poses a real danger to space travelers.

THE ULTIMATE MAP ▶

Want to find a stolen car, detect an oil leak in an ocean tanker, or determine the quickest route to your destination? The *Global Positioning System* (GPS) can do it for you. This system of about 24 satellites can locate anything on Earth within 20 to 300 yards.

A VISIT TO THE SUN

Although it's an average-sized star, the Sun is the largest object in the solar system. It contains about 98 percent of all the solar system's solid material. It's so big, more than one million Earths could fit inside. On the surface, the Sun is about 14,000°F

SPOTS AND FLARES

About every 11 years, the Sun develops an unusual number of spots. These dark patches on the Sun's surface are thousands of degrees cooler than the area surrounding them. Near the spots, huge columns of gas called *solar prominences* shoot out dramatically above the surface. Thin columns are called *solar flares*.

▼AURORAS

Sometimes, colorful, shimmering curtains of light fill the night sky near the North and South poles. These lights, called auroras, are caused by radiation from solar flares. Attracted to the magnetism of the Earth, the radiation spirals toward the poles as it approaches.

During a solar eclipse, the Moon passes between the Earth and Sun, blocking all but the outer rim from view.

SOLAR WIND

The Sun is like a huge nuclear furnace, producing incredible amounts of energy and radiation. Blowing out from the *corona*—the outer layer of the Sun—is a stream of particles known as *solar wind*. The wind gusts at 450,000 to 2 million mph, to the far reaches of the solar system.

92

▼ULYSSES

The probe *Ulysses* was sent off by the *Discovery* shuttle in 1990 to study the Sun. It was the first probe to study the solar system's so-called "third dimension"—above and below the plane in which the planets orbit the Sun.

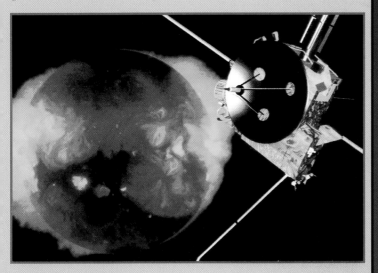

WHOA, H₂0!

Water on the Sun? Believe it or not, in May, 1995, scientists found enough water vapor in a sunspot to fill a lake four square miles in area and 300 feet deep!

SHOOTING STARS

Sometimes pieces of a comet break off. If they enter the Earth's atmosphere, friction causes them to heat up and burn. They are then known as meteors, but are often called "shooting stars."

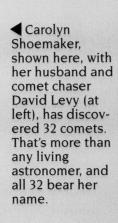

◀ Carolyn Shoemaker, shown here, with her husband and comet chaser David Levy (at left), has discovered 32 comets. That's more than any living astronomer, and all 32 bear her name.

93

▲ Halley's comet appears in our skies every 76 years and is due to return again in 2061.

DIRTY SNOWBALLS

Made of ice, rock, gas, and dust, *comets* are often called dirty snowballs. They move in orbits that take them close to the Sun, then back to the far reaches of the solar system. As a comet approaches the Sun, the ice vaporizes, and dust and gas are released.

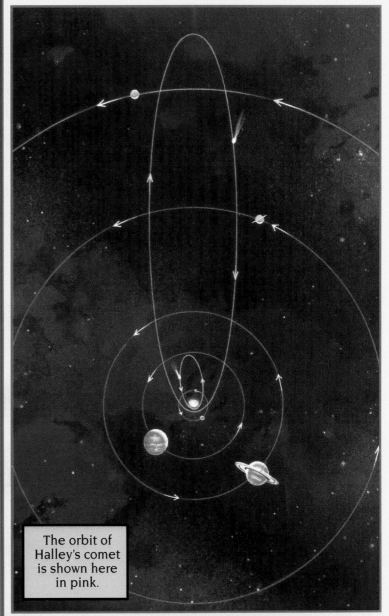

The orbit of Halley's comet is shown here in pink.

THE INNER PLANETS

Between the Earth and the Sun are the only two planets in our solar system that don't have moons. These inner planets are Mercury and Venus. And they are hot! Venus is the hottest planet in our solar system.

FLY-BY SNAPSHOTS

So close to the Sun, Mercury is a difficult body to observe through a telescope. But in March 1974, *Mariner* 10 flew by the planet and photographed its surface. Small, heavily cratered, and airless, Mercury resembles our own Moon. One crater on Mercury is 800 miles wide.

A GODDESS

Venus is named for the Roman goddess of beauty. It is a near twin of the Earth in terms of size. Except for the Sun and Moon, Venus is the brightest object in our sky.

A GOD ▲

The innermost planet, Mercury takes only 88 days to journey around the Sun. Because Mercury appears to Earthlings to be moving very quickly in its orbit, it is named for the Roman wing-footed god.

HIGHS AND LOWS

Without the protection of a gaseous atmosphere, Mercury gets hit hard by the Sun. The temperature can be as hot as 805°F. But the heat is lost at night, as there is no atmosphere to contain it. Then the temperature drops to about -275°F.

OUT OF SIGHT

Also like Earth, Venus is completely covered with clouds. It has an atmosphere. But the air on Venus is mostly carbon dioxide, which would suffocate humans. The cloud cover is 15 miles thick and yellow with sulfuric acid. Like a greenhouse, the clouds trap the Sun's heat and cause temperatures to hang around 900°F.

The *Magellan* probe.▶

Magellan took these photos of Venus as it orbited the planet.

▼ PLANET PEAKS

Spaceship *Magellan* was launched into space by a shuttle in May 1989. It arrived at Venus over a year later and revealed a rugged terrain. The Maxwell Mountains rise to more than 35,000 feet and may be the rim of an ancient volcano. Below, they are shown scientifically color enhanced.

BACKWARDS SPIN ▼

In December 1962, the U.S. spaceship *Mariner* 2 (below) traveled within 21,600 miles of Venus. The exploration confirmed that the planet takes 243 Earth days to *rotate*, or make a complete turn. And it rotates in the direction opposite to Earth—east to west.

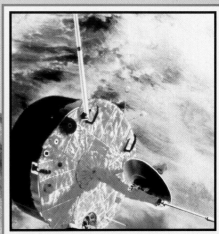

▲ Lava flows on Venus, photographed by *Magellan*.

VENUSIAN LIGHTS

Johann Schroeter, who lived in the late 18th and early 19th centuries, was the first to see a real phenomenon on Venus. The "ashen light" that he observed was thought to be the city lights of a Venusian civilization. It is now thought to be lightning.

95

THE RED PLANET

The rusty-red surface of Mars reminded many early observers of a bloody battlefield. The Romans named it after their god of war, and the name is still used today. However, we now know that Mars is red because its soil contains iron that has rusted.

◀ In 1938, actor Orson Welles created a panic when his radio broadcast of *War of the Worlds* was thought to be real news.

MONSTER MOUNTAIN

Although there are no active volcanoes on Mars, the sleeping ones are the tallest yet discovered in our solar system. Olympus Mons rises about 17 miles. That's more than 3 times as tall as Earth's tallest mountain—Mount Everest!

COMMON CHARACTER

Mars has more in common with Earth than does any other planet. Although the Martian year is almost twice that of Earth's, the Martian day is only 41 minutes longer than an Earth day. Also, Mars has four seasons. It has no running water, but Mars does have polar ice caps.

THE ALIEN THREAT

We tend to think of Martians when we think of Mars. Maybe that's because of the science-fiction novel *War of the Worlds*. Written by H.G. Wells in the 19th century, the novel tells of super-intelligent Martians invading the Earth with powerful fighting machines.

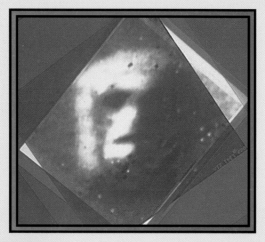

◀ In a photo of Mars's surface, people noticed a strange monkey-face pattern. It reminded them of "the man in the Moon."

VIKING VICTORY ▶

In 1975, the U.S. launched two *Viking* spacecraft toward Mars. Each had an orbiting module and a landing module. The two *Viking* landers set down on Mars; the *Viking* 1 *Orbiter* studied the Martian moon called Phobos; and the *Viking* 2 *Orbiter* surveyed Mars's second moon, called Deimos.

▼ *Viking* preparing to launch.

MAPPING MARS

The first spacecraft to pass near Mars, between 1965 and 1969, collected data that showed a cratered and lifeless surface. In 1971, the *Mariner* 9 spacecraft mapped the whole planet, revealing a great network of waterless riverbeds and huge volcanoes.

▼ An asteroid.

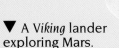

▼ A *Viking* lander exploring Mars.

THE BELT BEYOND

No Martians have visited Earth, but asteroids have. Beyond Mars is the asteroid belt. As large as 600 miles wide, these asteroids orbit the Sun, sometimes crossing Earth's path. If they enter Earth's atmosphere, they are known as *meteors*. If they don't burn up in the atmosphere, they may hit the planet's surface and cause craters.

The largest crater on Earth made by an asteroid is in Arizona. It's 4,150 feet in diameter and about 575 feet deep.

TWO GAS GIANTS

Jupiter and Saturn are the most gigantic planets in our solar system, and they are made mainly of gases. When observing them from space, we see only their atmospheres. But unmanned probes are learning more and more about them.

ROUGH ENTRY ▲

In December 1995, the *Galileo* probe became the first object from Earth to enter Jupiter's atmosphere. After a six-year flight from Earth, *Galileo* penetrated at over 100,000 miles per hour. Before it melted, the unmanned probe relayed a 57-minute weather report to NASA.

From front to back of photo are ▶ Jupiter's planet-sized moons Callistro, Ganymede, Europa, and Io (not shown to scale).

◀ COMET CRASH

When Shoemaker-Levy 9, the comet, passed too close to Jupiter in 1993, the planet's huge gravitational force pulled it even closer. The comet was torn to pieces. About 20 chunks remained in orbit around Jupiter until 1994, when they crashed into the planet. Each crash left a scar on Jupiter.

MANY MOONS

Both Jupiter and Saturn are orbited by many moons—16 are known to Jupiter, and 20 to Saturn. Jupiter's Io is the only moon in our solar system known to have active volcanoes. Ganymede is the largest moon in the solar system.

KING PLANET

The largest planet in our solar system is named for the king of all Roman gods, Jupiter. It's so big that 1,317 Earths could fit inside it. Its gravity is also much greater than Earth's: If you weighed 100 pounds on Earth, you'd weigh 254 pounds on Jupiter.

◄ STRIPES, SPOTS, AND RINGS

The clouds in Jupiter's atmosphere give it a striped appearance. Among the stripes is a huge revolving storm known as the *Great Red Spot*. The spot is three times the size of Earth and was first observed about 300 years ago! The other visible storms, known as the *White Ovals*, formed in about 1940. When the U.S. *Voyager* spacecraft visited Jupiter in 1979, it discovered rings around the planet. Visible only from the dark side, away from the Sun, the two rings are composed of dark grains of sand and dust.

▲ The *Voyager* spacecraft flew by Jupiter and Saturn, and on to the outermost planets before leaving the solar system.

KING OF RINGS ►

Saturn is the outermost planet visible from Earth with the naked eye. Its most famous feature is its ring system. In 1610, the astronomer Galileo was the first to observe them. There are seven main rings, which are made of ice particles. There are thousands of smaller rings as well, discovered by *Voyager* 1 in 1980.

◄ BIG DADDY

In Roman mythology, Saturn is the father of Jupiter. As a planet, it is nearly as large as Jupiter. But Saturn is not as weighty, having only one-fourth the mass.

THE OUTER LIMITS

The most distant objects in our solar system were unknown to ancient astronomers. They were too far away to see. It took better and better telescopes to find the planets in the outer limits. Uranus was sighted in 1781, Neptune in 1846, and Pluto in 1930.

◀ COOL BLUE, RINGS TRUE

A gaseous planet, Uranus is blue-green in appearance and very cool, with atmospheric temperatures as low as -366°F. According to *Voyager* 2, which flew by Uranus in 1986, the planet also has rings. The outer edge of the ring system is located 15,800 miles from the planet's cloud tops.

SIDE ROLL

Being tilted in orbit is not uncommon for planets, but Uranus is almost completely on its side. That means its poles, rather than its equator, point alternately toward the Sun.

SPOT ▶ AND SCOOTER

A storm rages in the atmosphere of Neptune. This blue planet once had a Great Dark Spot similar to Jupiter's Great Red Spot. But it has recently disappeared. Now another dark spot has appeared to the north, and to the south is a bright cloud, nicknamed "Scooter" because it moves so fast.

▲ Neptune also has rings, verified by *Voyager* 2 in 1989.

PLANET X

Around the turn of the 19th century, astronomer Percival Lowell began looking for what he called "Planet X," far distant in the solar system. He never found X. But, in 1930, 24-year-old Clyde Tombaugh (right) identified the new planet, which became known as Pluto.

PLUTO EXPRESS

The Hubble Space Telescope has obtained the first clear images of Pluto. But NASA is planning on obtaining even better pictures. In a mission called the Pluto Express, a pair of small spacecraft will study Pluto after the year 2000.

▼Astronomical observatories

▼ HIGH-TECH EYES

Telescopes were the first tools for exploring the heavens. They have really advanced since the early years—with gigantic mirrors and laser technology. Large telescopes in observatories are beginning to capture sharp pictures of objects in outer space.

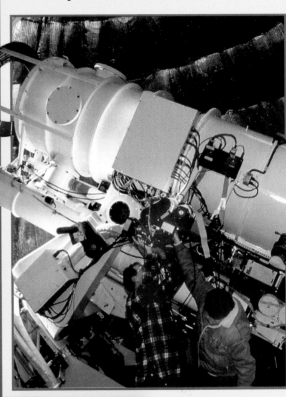

◄ CROSSING OVER

Pluto orbits the Sun every 248 Earth years. Having a more oval-shaped orbit, it periodically moves inside Neptune's orbit. It crossed in 1979 and will remain there until 1999.

◄ From left, Uranus, Pluto, and Neptune

DEEP SPACE

Our galaxy contains as many as 100 billion stars! Like our Sun, they are large, fiery balls of gas. But they don't burn forever. At their hottest, they are blue stars, but eventually they begin to cool down. Stars like our Sun, which is about 5 billion years old, may burn for about 10 billion years.

▼ STAR BIRTH

Stars come in many sizes and live all kinds of lives. They are born in a great cloud of dust and gas called a *nebula*.

From a nebula, stars arise. Gravity tugs at parts of the nebula, and globules develop. These blobs begin to spin faster and faster. As the center reaches about 15 million degrees Fahrenheit, the blob begins to shine and becomes a true star.

GUIDING LIGHT

The stars in our galaxy and the patterns they make stay constant in relation to each other. Since ancient times, they've helped sailors and other travelers navigate. Ancient skywatchers also saw pictures in the patterns of stars. These pictures are called constellations. Today, stars are grouped into 88 constellations. They are used by astronomers to memorize the positions of stars.

The constellation Centaurus.

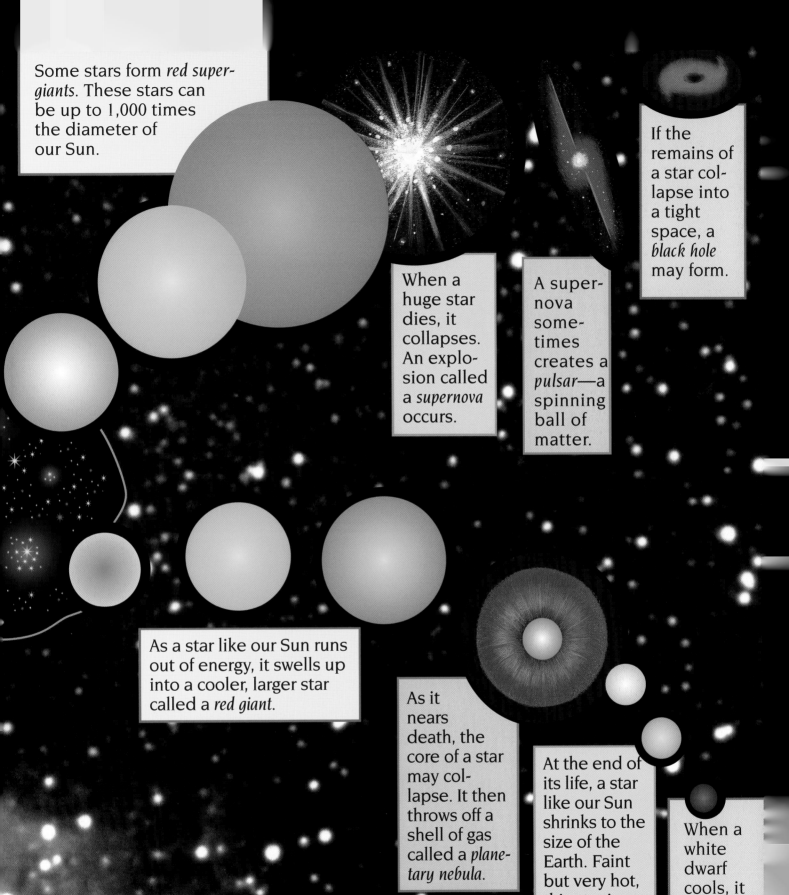

Some stars form *red super-giants*. These stars can be up to 1,000 times the diameter of our Sun.

When a huge star dies, it collapses. An explosion called a *supernova* occurs.

A supernova sometimes creates a *pulsar*—a spinning ball of matter.

If the remains of a star collapse into a tight space, a *black hole* may form.

As a star like our Sun runs out of energy, it swells up into a cooler, larger star called a *red giant*.

As it nears death, the core of a star may collapse. It then throws off a shell of gas called a *planetary nebula*.

At the end of its life, a star like our Sun shrinks to the size of the Earth. Faint but very hot, this star is called a *white dwarf*.

When a white dwarf cools, it becomes a *black dwarf*.

103

IMAGINATION AND BEYOND

Imagination has always fueled the study and exploration of space. Today, scientists are dreaming of a Moon station, planning trips to Mars and the distant solar system, and searching for life on other planets.

▲ Russian and American astronauts on the *Mir*.

WORLD MISSIONS

Since 1984, the United States, Canada, Japan, and the European Space Agency have been working toward putting a new space station into orbit. The size of two football fields, this outpost will orbit Earth every 90 minutes with a crew of six.

◄ Astronauts from the U.S. have docked their shuttle on the Russian space station known as *Mir*. The name *Mir* means "peace," or "community living in harmony."

ROBOT TO MARS

It weighs only 23 pounds and is about as big as a microwave oven. It's a little robot known as Rocky, and it's expected to answer some big questions about Mars. NASA's robot will analyze the rock and soil on Mars and possibly reveal whether life has ever existed there.

The Russians have developed a Mars robot, too.

NEW PLANETS

Other solar systems have been found at last! One has an ordinary star like our Sun, called Star 51. Orbiting around it is a Jupiter-sized planet. However, it's unlikely the planet has any life. It's positioned so closely to its sun that it probably sizzles at about 1,800° F.

This diagram compares part of our solar system to the three newly discovered solar systems. It shows how close the planets are to their sun.

OUR SOLAR SYSTEM

MERCURY   VENUS   EARTH   MARS

51 Peg Solar System

70 Vir Solar System

47 UMa Solar System

GALAXIES GALORE

Edwin Hubble established in the 1920s that our Milky Way is not the only galaxy in the universe. Pictures from the Hubble Space Telescope, which is named for the famous astronomer, have pushed the galaxy count from 10 billion to about 50 billion.

ADVENTURE ON EARTH

Most people don't get a chance to travel beyond Earth. But space adventures can be found right here on the ground. Look for comets passing overhead, like the Hyakutake, which passed over in March 1996. Watch for shooting stars, the phases of the Moon, or a solar eclipse. Space is right overhead!

Imagination rules at this celebration in Hollywood, where a young girl got a chance to meet Neelix and Kazon, characters from *Star Trek Voyager*.

Pioneer 10 was the first spacecraft to leave our solar system.

▲ MESSAGE AWAY

Alien characters often appear in TV and the movies, but do scientists believe that they really exist? Some do. Aboard the two *Voyager* spacecraft sent into space in 1977, there were messages describing life on Earth. Scientists hoped that the two ships might encounter alien civilizations. Who knows? Perhaps someday they will!

Index

Glossary

Acupuncture: Medical procedure in which needles are inserted into certain points of the body.

Anthropologist: Scientist who studies humankind.

Aquaculture: The breeding of fish in pens for human consumption; similar to a farm.

Aqua Lung: Breathing device, consisting of air-filled tanks, that allows people to explore the deep sea.

Aquanaut: Scientist who spends a long period of time living and studying underwater.

Archaeologist: Scientist who studies past cultures.

Atoll: Ring of coral that forms around a sunken volcano.

Auroras: Beautiful twinkling lights, caused by radiation from the Sun's energy, which can be seen from the northern and southern polar regions.

Bathyscape: Vehicle that explores very deep parts of the ocean.

Cluster: Fifth state of matter; has electrical, optical, and magnetic characteristics.

Comet: Frozen mass of rocky debris from space that orbits around the Sun.

Constellation: Group of stars seen from Earth and identified by name, such as Orion and the Big Dipper.

Corona: Outer atmosphere of the Sun.

Dust Devil: Swirl of dirt created when a tornado travels through the desert.

Earthquake: Shaking of the ground caused by sudden shifts of rock beneath Earth's surface along a fault.

Equator: Imaginary circle dividing Earth into two equal parts; the equator is the hottest place on Earth.

Food Chain: Series of living things in which each feeds upon the one below it and in turn is eaten by the one above it; cycle repeats itself until the tiniest animal eats the bacteria that is left behind from the largest animal.

Fujita Scale: Measures the strength of a tornado.

Galaxy: Cluster of stars held together by gravitational force.

Gale: Strong wind, not quite a tornado or hurricane, that blows at speeds of more than 32 miles an hour.

Gravity: Magnetic pull; force of attraction exerted by an object with mass.

Hurricane: Violent windstorm; also called a typhoon and cyclone.

Landslide: Quick downward movement of land caused by excessive rain or an earthquake.

Magma: Liquid rock inside Earth.

Marine Biologist: Scientist who studies the oceans.

Meteor: Mass of rock that enters Earth's atmosphere at high speeds.

Mollusk: Animal that has a soft, muscular body but no backbone; snails, mussels, and clams are mollusks.

Nanotechnology: Science of building tiny things the size of molecules.

Orbit: Path of Earth or other planets around the Sun.

Paleontologist: Scientist who studies fossils.

Plague: Serious and contagious illness that affects many people in a short period of time.

Richter Scale: Scale used to indicate the strength of an earthquake; also called a seismograph.

Robotics: Science of designing machines that work like living things.

Satellites: Natural or man-made objects in space that orbit a host planet.

Solar Eclipse: Occurs when a full or new Moon lines up directly between Earth and the Sun; only the outer rim of the Sun is visible from Earth during a solar eclipse.

Solar Flares: Thin columns of gas that shoot outward from the surface of the Sun.

Solar Prominences: Huge columns of gas that shoot outward from the surface of the Sun.

Solar System: Combination of planets, comets, asteroids, and all celestial bodies that revolve around the Sun.

Solar Wind: Particles that come from the outer layer of the Sun.

Spawn: To produce eggs.

Submersible: Small submarine that is used for research.

Tornado: Destructive storm, also called a twister, caused by thunderstorms.

Tsunami: Large tidal wave cause by an underwater volcano or earthquake.

Volcano: An opening in the surface of Earth, or any other planet, through which steam or liquid rock escape; once the liquid rock, or magma, flows on to the surface, it is called lava. Hardened lava builds up around the vent to form a cone-shaped volcanic mountain.